T0251872

Water Policy and Governanc in South Asia

Dr. Hossen carried out an exceptional program of research in Bangladesh which focused on water governance in relation to human rights, international water law and environmental sustainability. His major argument is that eco-agricultural system encounters major disruptions due to a number of factors including regional hydropolitics and neoliberal and highly centralized approaches to water resource management that follow the principles of "ecocracy." In this context, Dr. Hossen explores three major questions: (i) How can local ecological knowledge be incorporated into national water policy? (ii) What strategies and reforms are required at the international watershed governance level? and (iii) How can human rights principles, including the principle of water as a human right, be used to formulate more effective water policy and governance?

To answer these questions, Dr. Hossen explores the effects of regional hydropolitics on water management, focusing on three large engineering projects, the Farakka Barrage built by India on the Ganges River, and the Ganges-Kobodak (GK) Project and Gorai River Restoration Project (GRRP) Project in Bangladesh. To find out the effects of these projects, Dr. Hossen used his PhD research with one year fieldwork in 2011–12 based on focus group discussion, in-depth case study and social survey methods. In addition to this primary data, he looks at extensive secondary documents from the government of Bangladesh and research organizations pertaining to water management, agricultural modernization and institutional structures. The arguments herein are applicable particularly to the Ganges-Brahmaputra Basin countries in South Asia but also to the river basins of other parts of the world.

M. Anwar Hossen (PhD, UBC, Canada) is Professor of Sociology at Dhaka University, Bangladesh.

Routledge Studies on Asia in the World

Routledge Studies on Asia in the World will be an authoritative source of knowledge on Asia studying a variety of cultural, economic, environmental, legal, political, religious, security and social questions, addressed from an Asian perspective. We aim to foster a deeper understanding of the domestic and regional complexities which accompany the dynamic shifts in the global economic, political and security landscape towards Asia and their repercussions for the world at large. We're looking for scholars and practitioners – Asian and Western alike – from various social science disciplines and fields to engage in testing existing models which explain such dramatic transformation and to formulate new theories that can accommodate the specific political, cultural and developmental context of Asia's diverse societies. We welcome both monographs and collective volumes which explore the new roles, rights and responsibilities of Asian nations in shaping today's interconnected and globalized world in their own right.

The Series is advised and edited by Matthias Vanhullebusch and Ji Weidong of Shanghai Jiao Tong University.

Water Policy and Governance in South Asia

Empowering Rural Communities

M. Anwar Hossen

Routledge
Taylor & Francis Group

LONDON AND NEW YORK

First published 2017 by Routledge

2 Park Square, Milton Park, Abingdon, Oxfordshire OX14 4RN

711 Third Avenue, New York, NY 10017

Routledge is an imprint of the Taylor & Francis Group, an informa business

First issued in paperback 2018

Copyright © 2017 M. Anwar Hossen

The right of M. Anwar Hossen to be identified as author of this
work has been asserted by him in accordance with sections 77 and
78 of the Copyright, Designs and Patents Act 1988.

All rights reserved. No part of this book may be reprinted or
reproduced or utilised in any form or by any electronic,
mechanical, or other means, now known or hereafter invented,
including photocopying and recording, or in any information
storage or retrieval system, without permission in writing from the
publishers.

Notice:
Product or corporate names may be trademarks or registered
trademarks, and are used only for identification and explanation
without intent to infringe.

British Library Cataloguing-in-Publication Data
A catalogue record for this book is available from the British Library

Library of Congress Cataloging-in-Publication Data
A catalog record for this book has been requested

ISBN: 978-1-138-69066-0 (hbk)
ISBN: 978-1-138-36740-1 (pbk)

Typeset in Galliard
by Apex CoVantage, LLC

Water policy and governance in South Asia

Empowering Rural Communities

Water policy and governance is currently the central point of protecting river bank community's livelihoods as river water resources are transferring from local community specific to uniform management system. Local communities are mostly agricultural based and their traditional farming practices are deeply embedded in local culture, history and ecology. In this context, the Ganges River water based on the agro-ecosystems is everything to local communities, natural environment, ecosystem health and human rights in the Ganges Dependent Area (GDA) in Bangladesh. However this system encounters major ecocracies due to hydropolitics in the Ganges Basin countries and power structure in the global and national levels. Due to the Farakka diversion, for example, the GDA communities encounter major challenges for accessing ecological resources like cropland siltation, wild vegetables, vegetation, local seeds and irrigation water. Based on supports of global political economy, the government in Bangladesh established water and agricultural management systems to overcome the effects of Farakka diversion; however, because of top-down nature of these systems, the marginalized farmers face further survival challenges. Those challenges result in human rights abuses, since they mean that many households are now unable to secure basic human rights, as defined by the United Nations (1948), to food, water, education, housing and health care. Based on the UN (2002) definition of access to water as human rights, my argument is that the Ganges River water right and ecosystems are the prerequisite for maintaining these human rights. Only water governance at local, national and bain-wide levels can overcome the current challenges and can protect the human rights. Based on this argument, this study employed ethnographic methods, informed by the fields of environmental anthropology and political ecology, to gather information about traditional farming practices, water management, local ecological knowledge, local economy and human rights issues.

M. Anwar Hossen is Professor of Sociology at University of Dhaka, Bangladesh. He is co-editor of Springer Journal, Bandung: Journal of the Global South. He published an edited volume in 2016 on Water and Ecological Resource Governance as Guest Editor with this journal. Dr. Hossen has extensively published research articles at international level (e.g., Asiatic Society, Brill, Routledge and Springer) and at national level (e.g., Social Science Review). He is the recipient of Nehru Humanitarian Award in 2010 from the University of British Columbia, Canada and of doctoral research award in 2011 from International Development Research Center (IDRC) Canada.

To my parents
M. Abul Kasem and Mrs Ayesha Kasem

Contents

Figures

Tables

Maps

Preface and acknowledgements

This book is developed based on my PhD dissertation at the University of British Columbia (UBC) Canada. The main foundation of this dissertation argument; marginalized communities' human rights depend on the river water right, originated from my research at UBC and Carleton University as well as my personal backgrounds. During my MA degree of Sociology and Anthropology department at Carleton, I did a research on Natural Disaster, Inequality, and Vulnerability: A Case Study in Rural Bangladesh. In this research, I explore the issue of *access to resources (or the lack of access to resources)* as a critical factor in allowing some to survive the flooding with little damage and others to be devastated. At the end of this research, I realized that the basin hydropolitics as well as global power structure are bigger factors than local dynamics to understand this access issue. Therefore, I determined to begin my PhD and the UBC accepted me as a student. During this study, I was fortunate to connect my boyhood's understanding of ecosystem and ecological resources to develop my research argument. I was born and raised in the Ganges-Brahmaputra Basin (GBB) confluence point at Manikgonj District in Bangladesh where we received enormous ecological resources. I was also very happy to experience my parent's sharing of surplus resources with neighboring marginalized people. Rather than my birth place at the GBB, I selected my PhD fieldwork site, Chapra at Kushtia District, on the Ganges Basin for the three major reasons. Firstly, Chapra is situated on the Gorai River, a branch of the Ganges River in Kushtia, and faces major ecological effects directly from the Farakka diversion. Secondly, this site is currently dependent on the Ganges-Kobodak (GK) project, the largest irrigation project for agricultural production in Bangladesh. Thirdly, Chapra village is five feet away from the Gorai River Restoration Project (GRRP) that represents one example of water management approach in Bangladesh.

As this book sheds lights on my personal and academic backgrounds, it is important to acknowledge my parents and the people who teach me how to respect nature, community and social justice. During my studies at Department of Sociology, Dhaka University–Bangladesh, Professor S. Aminul Islam encouraged me to study the impacts of recurring floods on human behavior. At Carleton, Jared Keil as supervisor of my MA thesis at Sociology and Anthropology Department was helpful to promote my academic understanding of this issue.

At UBC, I am very grateful to John Wagner who was so kind to accept me as his PhD student. Without his academic and funding supports, it was very difficult to complete my PhD. Furthermore, I must acknowledge to Naomi McPherson for her pedagogical supports during my study-in-progress. Again, I am very grateful to the International Development Research Centre (IDRC), Ottawa for offering me a Doctoral Research Award for doing my fieldwork in Bangladesh. I am also grateful to the UBC for awarding me the Nehru Humanitarian Award in 2010 and for funding supports throughout my PhD studies.

I would also like to thank Dr. Matthias Vanhullebusch for his invaluable suggestions during the book proposal stage. Furthermore, I was fortunate to get critical comments on my dissertation from Craig Jenkins, Professor of Sociology, Political Science & Environmental Science, The Ohio State University, and Rezaur Rahman, Professor of IWFM, BUET. ShengBin Tan, editorial assistant of Routledge, was very helpful to make my tasks easier during the review processes of my dissertation.

Acronyms

ASK	Ain O Salish Kendra
ASP	Ammonium Sulfate Phosphate
BADC	Bangladesh Agricultural Development Corporation
BARC	Bangladesh Agricultural Research Council
BARD	Bangladesh Academy for Rural Development
BBS	Bangladesh Bureau of Statistics
BDT	Bangladesh Taka
BFA	Bangladesh Fertilizer Association
BFRI	Bangladesh Fishery Research Institute
BLAST	Bangladesh Legal Aid and Services Trust
BRDB	Bangladesh Rural Development Board
BSEHR	Bangladesh Society for the Enforcement of Human Rights
BSF	Border Security Force
BWDB	Bangladesh Water Development Board
DAE	Department of Agricultural Extension
DAP	Diammonium Phosphate
DoE	Department of Environment
DoF	Department of Fisheries
EGPP	Employment Generation Program for the Poorest
FAP	Flood Action Plan
FCDI	Flood Control, Drainage and Irrigation
FFW	Food For Work (KABIKHA)
GBB	Ganges-Brahmaputra Basin
GDA	Ganges Dependent Area
GK	Ganges-Kobodak (Project)
GR	Gratuity Relief
GRRP	Gorai River Restoration Project
HYV	High Yielding Variety
IFC	International Farakka Committee
IMF	International Monetary Fund
IWRM	Integrated Water Resource Management
JRC	Joint River Commission
KCERP	Khulna Coastal Embankment Rehabilitation Project

KIDRP   Khulna-Jessore Drainage Rehabilitation Project
LGED   Local Government and Engineering Department
LGRD   Local Government for Rural Development
MDGs   Millennium Development Goals
MoA   Ministry of Agriculture
MoF   Ministry of Finance
MoDMR   Ministry of Disaster Management and Relief
MoFDM   Ministry of Food and Disaster Management
MoHFW   Ministry of Health and Family Welfare
MoWR   Ministry of Water Resources
MP   Member of Parliament
NGO   Non-Government Organization
NRLP   National River Linkage Project
NWMP   National Water Management Plan
NWP   National Water Policy
PDB   Power Development Board
PIC   Project Implementation Committee
PRSP   Poverty Reduction Strategy Paper
RAB   Rapid Action Battalion
SSP   Single Super Phosphate
TR   Test Relief
TSP   Triple Super Phosphate
UBINIG   Unnayan Bikalper Nitinirdharoni Gobeshona
    (Policy Research for Development Alternatives)
UC   Union Council
UHC   Upazila Health Complex
UNO   Upazila Nirbahi Officer
UP   Upazila Parishad
VGD   Vulnerable Group Development
VGF   Vulnerable Group Feeding
WARPO   Water Resource Planning Organization

Glossary of Bengali terms

Ain O Salish Kendra	name of a human rights NGO
Aman	the most important local rice variety grown during the wet season
Andalon	social movement
Ashrayan	a government housing program; the literal meaning of the name is "place of shelter"
Aus	the most important rice variety grown at the end of the Spring season
Baacha morar lorai	do or die struggle
Beel	wetland
Beel Dakatia Andolon	Dakatia Wetlands Movement; a local movement in Khulna District
Bigha	a unit of land measurement; in Kushtia District 1 bigha equals 0.3 acres
Bhukha michhil	demonstration against food shortage
Bonna	extreme flood
Bonsher baati	line of inheritance
Boro	name of a rice variety now commonly grown in place of aus and aman due to continuous ecosystem failures
Bosenter kokil	a spring bird
Charland	sand accumulation in a river bed
Chashi	traditional farmer
Chor	thief
Dadagiree	an elder brother's authoritative practice
Deki	paddy husking pedal
Dharmaghat	strike or withdrawal of service as a form of protest
Dhubba	grasses that grow in between rows of crops
Dhup	fragrance of a particular herb
Ejara	a lease
Gajar	snakehead fish
Gucchagram	a government housing program; the literal meaning of the name is "cluster of houses"

Eid-ul-Azha	Muslim festival of sacrifice celebrated on the tenth day of the month of Jilhajj in the Bengali calendar
Eid-ul-Fitr	Muslim festival on the first day of the month of Shawal to celebrate the end of Ramadan (fasting)
Gom	wheat
Gorur Garee	bullock cart
Ha-do-do	a Bengali community outdoor sport
Haor	oxbow lake
Harijon	scheduled caste
Harichaada	traditional toll collection
Hartal	a street demonstration that involves the closure of shops, offices and transportation systems
Imam	Muslim preacher
Jaigidar	food and accommodation in exchange for private tutoring
Jee hujur	yes sir
Joler dame	incredibly lower price
Joke	blood sucking worm
Kaachi	scissors
Kakla	crocodile fish
Kalijera	local name for rice paddy
Kamla	day laborer
Kata diyea kata tola	set a thief to catch another thief
Kecho	earthworm
Khal	a canal
Khal Khanon kormochuchi	a canal renovation program
Kharif 1	agricultural cropping period from February to April
Kharif 2	agricultural cropping period from May to August
Khora	dry season after winter
Kola bang	bull frog
Kolosh	earthenware jar
Kot	traditional economic transaction system
Krishak	modern farmer
Kuichya	eel
Kutcha	house built with straw and mud
Kula	winnow fan
Kuya	ring-well
Langol	traditional plow pulled by bullocks
Maacha	a raised platform made with bamboo and rope
Matobbar	a village chief
Mohajon	a traditional money lender
Nagorik Oikya	citizens' forum
Nerica-10	hybrid rice variety that requires a strictly regulated water supply

Nouka bice	boat race
Noya Krishak Andolon	new farmers' movement
Odhikar	name of a human rights NGO
Paisa	one unit of Bangladesh currency; 100 paisas equals one Bangladesh Taka
Pakhi	a unit of land measurement; one pakhi equals 0.6 acres
Paribesh Andolon	environmental movement
Pitha	traditional cake or sweetmeat
Polo bice	community fishing sport
Pucca	house made with brick and cement
Puja mandir	Hindu prayer house
Purdah	Muslim religious practice of seclusion according to which the female body is covered from head to toe
Puti	minnow
Rakhal	yearly contract laborer
Rakkhosh	monster
Robi	Agricultural cropping period from October to January
Sada sar	white fertilizer or urea
Semipucca	house made with brick, corrugated iron, and mud
Shaheb	a form of address to an adult male; Mr.
Shako	locally made bridge over a small river or canal
Shikor	root
Sonar Bangla	Golden Bengal
Songkot	crisis
Sorkarer Dalal	agent of government
Swanirvar	a government employment program for marginalized people's empowerment
Ufsi aus	hybrid rice variety that requires a strictly regulated water supply
Union Council	lowest level of local government that includes 12 wards made up of 20 to 30 villages
Unnoti	development
Upazila Parishad	sub-district level of local government
Vaate marbe paanite Marbe	local people will be killed by rice and water deprivation
Vater-maar	a scum or liquid substance made from boiled rice paddy
Vita	raised bed for growing crops
Zamindar	a landlord

Introduction

Agricultural communities in the Ganges Dependent Area (GDA) of Bangladesh are facing extreme hardship due to hydropolitics, global power structure and government's water and agricultural management approach. To understand this hardship, this book describes the effects of basin hydropolitics on the marginalized basin communities, focusing on three large engineering projects, the Farakka Barrage built by India on the Ganges River, and the Ganges-Kobodak (GK) project and Gorai River Restoration Project (GRRP) in Bangladesh. Based on these projects, I describe challenges of traditional livelihoods and local ecological knowledge of farming households in the community of Chapra, Kushtia District, Bangladesh, and the rapid displacement of traditional practices by a commoditized system created through government interventions. India is a major contributor to these hardships because of its diversion of major portion of water from the Ganges River before it reaches to Bangladesh. The government systems within both Bangladesh and India are best understood as ecocracies, that is, highly centralized and bureaucratic systems in which resources are controlled by elite groups following neoliberal development approach.

Based on this argument, the main goal of this book is to document the hardships currently being faced by agricultural communities in the Ganges Basin in Bangladesh. Another important goal is to develop policy recommendations to improve water governance within the Ganges Basin and thereby reduce the hardships of farming communities to loss of livelihoods. In order to determine what improvements to the governance system would be practical, this book addresses the following research questions: (i) how can local ecological knowledge be incorporated into national water policy in Bangladesh; (ii) what strategies and reforms are required at the international watershed governance level; and (iii) how can human rights principles, including the principle of water as a human right, be used to formulate more effective water policy and governance? In order to accomplish these three questions I acquired in-depth knowledge of the farming and livelihood practices at my fieldwork site, Chapra. This was done through a combination of participant-observation, interviews, focus group discussions, in-depth case studies and household surveys, all of which were carried out in 2011–12 in Chapra community in Bangladesh. I gathered information about the effects of the three projects, the Farakka, GK and GRRP, on this community in terms of agricultural production and human rights protection. In addition to this primary data collection, extensive secondary documents were collected from the government of Bangladesh and other agencies pertaining to water management.

My fieldwork data found that the Ganges River flow is currently very unpredictable, and there have been major ecosystem failures; sometimes this flow begins later or earlier and higher or lower due to the upstream intervention of Ganges Basin. The Farakka Barrage diversion in India dominates longer or shorter wet or dry season. The diversion is one cause of ecological vulnerabilities like flooding, drought and sedimentation on the Gorai River Bank communities which are responsible for agricultural practice challenges for Chapra people. Therefore, those farming households who constitute the vast majority of the population in the GDA now find themselves more and more vulnerable to human rights protection. This purpose of my book is therefore to document local practices and livelihood challenges and, on that basis, identify the changes in water policy and governance necessary to alleviate current hardships.

To overcome these hardships, water and agricultural management projects like the Gorai River Restoration Project (GRRP) and the Ganges-Kobodak (GK) Project can be described as major initiatives of the government in Bangladesh. These projects have some initial success in increasing crop production with improved water supply. However, these projects, in the long run, increase ecological vulnerabilities like river bank erosions and *charlands* at Chapra and other areas that cannot be overcome with further advanced technologies. The effects of one technology create demands for multiple technological involvements in overcoming the concerns. The Farakka diversion creates importance of the GRRP for removing sedimentations and *charlands*. However, the GRRP fails to overcome these problems and increases multiple risks of floods, water stagnations, embankment failures and river bank erosions. The failures of the Gorai River make Chapra villagers dependent on the GK project. At the initial stage, the project had success in promoting water supply for crop production. However, the project later on fails to provide this irrigation water and creates major challenges in water commodification and market economic domination.

In this context, it is worth noting here that a human rights approach introduces new and challenging questions into the governance debate. Given the hydropolitics of South Asia as a whole and the current neoliberal domination over local resource management throughout the region, is it feasible to think that a marriage (a healthy relationship) between ecosystems and economic development in the Ganges Basin could be achieved? A marriage of this type, at a minimum, requires greater respect for and incorporation of local ecological knowledge, and a coordinated, multi-scalar governance approach involving all the countries of the Ganges Basin. Without this fundamental change to existing policies, it will not be possible to secure human rights of the marginalized farming households living in the Ganges Basin. This governance approach is a major effort to promote community empowerment and sustainable development in Bangladesh.

Organization of the book

This book is comprised of eight chapters, developed to document the hardships currently being faced by agricultural communities on the Ganges Basin in Bangladesh, and to develop policy recommendations to improve water governance within the Ganges Basin and thereby reduce the vulnerabilities of farming communities to human rights abuses and loss of livelihoods in Bangladesh. Chapter one introduces theoretical approaches in the context of the Ganges Basin ecosystem, hydropolitics, economic development and community livelihoods. This chapter also describes research methods used to gather the data necessary to answer the research questions. In the second chapter, I describe the impacts of reduced flows of Ganges River water on the local ecosystem and on livelihood practices at Chapra. The fieldwork evidence confirms that the diversion of water by India at the Farakka Barrage point is reducing local ecological resources and increasing vulnerabilities to flooding, drought and riverbank erosion. This chapter also focuses on the way in which the political culture in Bangladesh causes further hardships for local communities by ignoring their voices in favor of a top-down approach that serves elite interests.

In order to fully document the effects of top-down water management in Bangladesh chapter three analyzes two major water management projects: the Gorai River Restoration Project (GRRP) and the Ganges-Kobodak (GK) Project. Using fieldwork data, I explain how and why the water development approach has failed to overcome the negative effects of the Farakka diversion and has contributed to additional ecological vulnerabilities of the Ganges Dependent Area (GDA) in Bangladesh. I also explore why the GK project has not been able to realize its goal of securing sufficient water supply for agricultural production in this area.

Currently, local people at Chapra are experiencing a radical transformation of their agricultural practices, which are described in chapter four. All aspects of the productive cycle, including the supply of seeds, fertilizer, land and water are subject to new forms of commodification controlled by state agencies and elites. The government of Bangladesh attempts to mitigate the hardships and social inequality caused by these forms of commodification through safety-net programs that subsidize the cost of new agricultural materials and technologies, or provide employment, housing, education and health benefits to the poorest households. In chapter five, I describe the role of local governments in distributing these safety-net program benefits and the extent to which these benefits are captured by local elites. Local governments fail to perform proper roles to protect the marginalized communities at Chapra due to top-down domination over the safety-net programs.

Chapter six describes the extent to which the hardships now being experienced by Chapra residents constitute human rights violations based on international customary law and a series of conventions and agreements created by the United Nations. The marginalized households experiencing these hardships are raising their voices and resisting their marginalization in many ways including participation in local ecological movements, which are described in chapter seven. These movements are occurring at local, national and international levels. In this context, chapter eight summarizes the data and returns to the three research questions described in the introduction. This chapter also formulates recommendations for how water governance in the Ganges Basin can be improved so as to protect the human rights of marginalized households and the critical ecological characteristics on which their livelihoods depend. This governance approach seeks to give voice to the marginalized farming households in the Ganges Basin for protecting their human rights.

1 River water, power and agricultural community livelihoods

The Ganges Basin context

Bangladesh is a delta country dominated by an agrarian society. Two of the largest rivers in Asia, the Ganges and Brahmaputra, meet in Bangladesh before emptying into the Bay of Bengal. The Ganges River and its major distributaries dominate the southwest region of the country, where this study has done. Rural communities in this region produce agricultural crops on a year round basis, following seasonal patterns and the variable flow of water in the Ganges River. However, this eco-agricultural system is now undergoing major disruptions due to a number of factors including regional hydropolitics, neoliberal and highly centralized approaches to water resource management that follow the principles of "ecocracy" as that term has been used by Escobar (1996) and Sachs (1992). The central governments of both India and Bangladesh follow much the same principles as they seek to manage natural resources like river water through a combination of engineering projects, bureaucratization and the application of economic liberalization policies. But, unfortunately for Bangladesh, India holds the upper hand when it comes sometimes to their competing interests over water resources. The central government in India, for instance, constructed the Farakka Barrage unilaterally on the Ganges River, very near the Bangladesh border, reducing flows to agricultural communities in Bangladesh at critical times in their cropping cycles. The barrage was built to divert water to Kolkata, to enhance navigation and shipping and to provide the city with fresh water. The central government in Bangladesh has been unable to restore this flow to a sufficient level and subsequently has intensified its own top-down water management approaches. The ethnographic fieldwork from 2011 to 2012 finds out effects of these approaches at Chapra in Bangladesh. The fieldwork data evince that these approaches fail to recognize the full range of ecosystem services on which most people rely and they encounter multiple survival challenges. To elaborate this argument, it is important at first to describe ecological characteristics of the Ganges Basin flow along with hydropolitics and theoretical approach.

Ecological characteristics of the Ganges Basin

The Ganges Basin is best understood as an ecosystem that is part of the larger Ganges-Brahmaputra Basin that originates in the Himalayan Mountains, flows through China, Nepal, India and Bangladesh, and ends in the Bay of Bengal (Map 1.1).

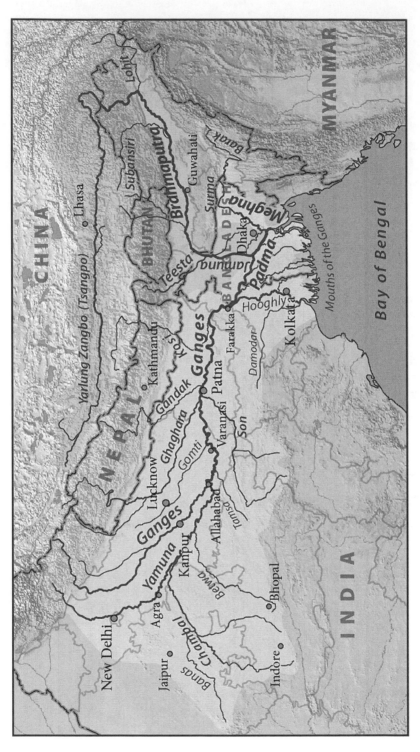

Map 1.1 The Ganges-Brahmaputra Basin in India, China, Nepal, Bhutan and Bangladesh

Source: Wikimedia (2013)

The Alakananda and Bhagirthi Rivers are the two major headwater sources of the Ganges where it arises in the western Himalayas in India at an elevation of 23,000 feet. After merging, these two rivers flow south to the Ganges Plain, north of New Delhi, and then continue eastward across the plain, traversing almost the entire country, India, before reaching Bangladesh. Along the way it is fed by several other major Himalayan tributaries, including the Yamuna, Gomati, Rama Ganga, Ghaghara, Gandak and Kosi rivers which originate in Nepal (Parua 2010:267–268). The headwaters of the Brahmaputra, on the other hand, arise on the northern side of the Himalayas and flow east first across the Tibetan Plateau, before turning south to India and then west to Bangladesh. The Ganges River in Bangladesh is known as the Padma and the Brahmaputra is known as the Jamuna, but throughout this book I use their more commonly recognized names.

The Ganges-Brahmaputra Basin in Bangladesh includes 57 international and 310 local rivers (Bangladesh Bureau of Statistics 2010a). About 94 percent of the water flow within these rivers in Bangladesh originates from outside the country in India (Elhance 1999:158). Most of Bangladesh is less than six meters above mean sea level (Gupta et al. 2005:387). This low elevation contributes to the fact that more than 1 million of the country's total 14 million hectare surface area is covered with local water bodies (Bangladesh Bureau of Statistics 2010a), like the Chapaigachi oxbow lakes, Shinda and Shaota canals, and the Lahineepara and Chapra wetlands at Chapra, which are replenished with water flow from the Gorai River, a distributary of the Ganges.

Agricultural practices at Chapra have developed over many centuries based on the river flow and the seasonal characteristics of the region. The two main seasons are known as *borsha* (wet season), which occurs during four months from June to September, and *khora* (dry season) which occurs between February and May, with some variation, depending on the volume of water in the Ganges River. These seasonal patterns occur within what is overall a moderately hot and humid climate, and this combination of climate, seasonal variation, low elevation, and abundant wetlands, provides the foundation for traditional agricultural and employment practices at Chapra.

Water in this setting, when analyzed in relation to social, economic and cultural practices, constitutes what Orlove and Caton, relying on Mauss (1990), have referred to as a "total social fact" (Orlove and Caton 2010:402). Every aspect of people's lives at Chapra is bound up with the movement of water through the environment and their ability to adapt to its seasonal characteristics. Traditionally, every household had egalitarian access to the flow of water during the borsha season for agricultural (Islam and Atkins 2007:131) and other uses (Zaman 1993:995). Water can thus be considered an essential actor within this socioecological system, possessing its own form of agency (Latour 1995; Wagner 2013). This book explores how the realization of fundamental human rights is dependent on the characteristics of this socioecological system and on the movement of water through it.

Currently, because the Ganges River flow is very unpredictable, there have been frequent ecosystem failures; some croplands face flooding or water stagnation for longer than normal periods in a year and other croplands, at other times of the year, face prolonged drought. River levels can no longer be predicted on the basis of natural seasonal changes but depend also on the mercy of the Farakka Barrage diversion. The diversion now determines whether wet and dry seasons will be longer or shorter than normal. These ecosystem changes create enormous vulnerabilities for most local farming households and are responsible for major agricultural and employment practice failures in the community of Chapra, my primary research site. The poorer farming households that constitute the vast majority of the population of the Chapra region and the Ganges Dependent Area (GDA) as a whole now find themselves more and more vulnerable to livelihood failures. These failures result in what can be considered as human rights abuses, since they mean that many households are now unable to secure basic human rights, as defined by the United Nations, to food, water, education, housing and health care. Based on the United Nations (2002) definition of access to water as human right, I argue in this book that the Ganges River water right, including the right to the ecosystem services it provides, is the prerequisite for maintaining other human rights.

The situation at Chapra is complicated by the fact that hardships are unevenly distributed. A small number of farming households are relatively well off by comparison to the majority and constitute an elite group with strong connections to government. These connections allow them to dominate local water management institutions and benefit from capital intensive agricultural programs. The elite farmers are market oriented and increasingly less dependent on the ecosystem services and on common property resources on which the majority depend. Therefore, water and agricultural management in Bangladesh adds new causes of ecosystem failures in addition to the Farakka diversion. To understand international dimension of these ecosystem failures, it is important to discuss here the hydropolitics of the Ganges Basin in South Asian Countries.

Hydropolitics in South Asia

Based on the international water laws and conventions, India, as a downstream country, resists water projects of the upstream China, Nepal or Bhutan. However, India as an upstream country does not follow this principle with its downstream Bangladesh (Brichieri-Colombi and Bradnock 2003:50; Elhance 1999:173). India's bilateral hydropolitics fails to recognize international water laws and conventions: the 1966 Helsinki Rules, the 1997 United Nations Watercourses Convention and the 2004 Berlin Rules. Recently, the government of China planned to generate hydropower by channeling the Tsangpo at Nuxia across the Great Bend. Moreover, China is also planning to augmentation of the Yellow River that is currently encountering water stress and about 800 kilometers away from the Tsangpo River (Brichieri-Colombi 2008:291). These Chinese water projects cause major concerns over the National River Linking Project

(NRLP) in India. China is at the upstream position and more powerful than India's. In this context, India needs to develop a coalition with the other countries of the Brahmaputra Basin to neutralize Chinese domination (Bhattarai 2009:5). This new realization can change India's bilateral approach and can promote multilateral approach (Hossen 2016).

India is the most rapidly developing and most water scarce nation in South Asia but also the most powerful politically and economically. Bangladesh suffers from this power difference since it is located downstream from India with respect to major shared rivers including the Ganges and Brahmaputra. Political conflict in the region is also an outcome of colonial legacies. British colonial policies laid the groundwork for the division of South Asia into the independent states that exist there now, and current political tensions in the region owe a great deal to this history. The British colonial legacy has also affected the ways in which modernization processes have unfolded and the way in which statecraft and governmentality have evolved (Pels 1997). In this setting as in many others, the colonial period has also strongly affected the relationship of contemporary governing elites to local communities (see for example Davidson-Harden et al. 2007; Escobar 1991; Ferguson 1994; Gardner and Lewis 1996:16; Kirby 2006:636; Langford 2005:274; Mehta 2005; Torry et al. 1979:523, 31; Turner 1997:13).

The British colonizers established a development discourse that described the colonized margins as backward, illiterate and traditional people in need of western style economic, educational and bureaucratic systems (Gardner and Lewis 1996:5). With the arrival of the postcolonial state, the people on the margins were ruled by the elite, bourgeoisie groups that had risen to power within the colonial bureaucracy or retained their authority as *Zamindars* (wealthy landowners) (e.g., Das and Poole 2004).

Large-scale water management projects were a fundamental aspect of the British colonial period. Technocentric irrigation systems in support of commercial agriculture were introduced to the Indian Subcontinent beginning with the East India Company in 1760s (Altinbilek 2002:10; Baviskar 2007:1; D'Souza 2006:621; Gilmartin 1994; Stone 1985; Verghese et al. 1994). In 1854 the British Government in India completed the construction of 2,298 miles of canals and distributaries in the upper Ganges alone (Khan 1996). D'Souza (2006:621) refers to these massive historical water interventions as "colonial hydrology" and the continuation of this approach today has been termed "hydrological nationalism" by Bandyopadhyay (2006:2). According to Bandyopadhyay, political leaders win elections on the basis of their ability to exploit basin hydrological resources in the name of national socioeconomic development. India, as an independent state, had built about 4,300 dams by 1994 and is considered one of the major dam building countries of the world (Hill 2006:149). This hydro-nationalism continues to inform India's approach to shared rivers, which is characterized by a strong emphasis on unilateral actions or, when agreements are reached with neighboring countries, on a bilateral rather than multilateral approach (Crow and Singh 2000:1910; Gyawali and Dixit 2001).

India utilizes a bilateral approach in order to limit the application of general principles that could work to its disadvantage (Elhance 1999:172; McGregor 2000). As a downstream country, for instance, India argues that upstream China, Nepal or Bhutan should not execute any action that can harm their interests as a downstream country. As part of their bilateral agreements with Nepal and Bhutan, India has funded the construction of several hydro-projects in Nepal and Bhutan that provide them with hydro-power at a very reasonable cost. However, upstream India built the Farakka Barrage on the Ganges River unilaterally, without consideration of the interests of downstream Bangladesh. They subsequently concluded a series of bilateral treaties with Bangladesh concerning the management of the Farakka Barrage and shared rivers more generally but did so on very unequal terms that fail to promote reasonable and equitable water sharing (Brichieri-Colombi and Bradnock 2003:53)

Despite the many well documented problems associated with the Indian approach to international waters, the government in India avoids participation in conflict resolution processes (Elhance 1999; Giordano et al. 2002; Gleick 2009; Haftendorn 2000; Wolf 1997), and they avoid invitations to develop a multilateral approach for basin-wide water management (Bandyopadhyay and Ghosh 2009; Brichieri-Colombi and Bradnock 2003; Gyawali and Dixit 2001). Indian hydro-domination in South Asia is thus increasing (Hossen 2012), with China as the only country in a position to challenge India's authority. The well documented effects of this particular approach to water management include loss of croplands, forest, fisheries, natural vegetation and habitat, due to increasing incidences of flooding, drought, river bank erosion, salinity and water stagnation (Bharati and Jayakody 2011; Hanchett 1997; Islam and Gnauck 2011; Islam and Karim 2005; Karim 2004; Mirza 1997; Mirza and Sarker 2004; Nilsson et al. 2005; Paul 1999; Potkin 2004; Swain 1996). The effects in Bangladesh are not caused only by the Farakka Barrage, but also by a series of other large barrages and structures that are capable of diverting 100,000 cubic feet per second (cusecs) from the Ganges flow before it reaches the Farakka point (Khan 1996). Currently, these water structures are used for commercial water and agricultural production (Altinbilek 2002:10; Baviskar 2007:1; Biswas and Tortajada 2001; Cox 1987; Khush 2001; Selby 2007; Van der Pijl 1998; Verghese et al. 1994).

Hydropolitics in Bangladesh

The effects in Bangladesh of the Farakka Barrage and other Ganges River diversions in India have been magnified by a series of actions taken within Bangladesh itself. As a result of flood damages experienced in 1954 and 1955, the Bangladesh Government, then East Pakistan, developed an agricultural modernization and water management plan with the goal of increasing cropland acreage by removing water from local wetlands and oxbow lakes (Alexander et al. 1998:1). The Flood Control, Drainage and Irrigation (FCDI) project, implemented in the early 1960s, with several sub-projects like the Ganges-Kobodak project in

Kushtia, which is described in detail in chapter three. FCDI projects were credited by some as worsening rather than preventing the floods of 1988 (Zaman 1993:989) and eventually FCDI was replaced by an even larger scale program of government intervention, the World Bank–sponsored Flood Action Plan. The Gorai River Restoration Project, which is described in more detail in chapter three, was implemented as part of the National Flood Control Plan and, like FCDI, is now criticized for having caused great hardship and ecological degradation and for being riddled with corrupt practices. This argument is supported my focus group respondents, Abonti, Gedi and Bimal: *odhik baad maane odhik moron faad* (the more embankments or dams mean the more death traps).

In coordination with its water development projects, the government in Bangladesh has also supported a series of agricultural development and food security projects beginning in 1960s (Amin 1974; Sobhan 1981:328). The Green Revolution was one major component of this modernization approach (Cleaver 1972:184), promoting new technologies like chemical fertilizers, pesticides, HYV crops and controlled irrigation. The Ford Foundation and other international agencies provided funding support for this revolution and the Bangladesh Academy for Rural Development (BARD) was established to promote this agricultural modernization (Blair 1978:65–66).

Theoretical approach

In this book, I rely on four closely interlinked bodies of theoretical literature in my approach to these issues. Literature from the field of development theory (Escobar 1996; Ferguson 1994; Gardner and Lewis 1996; Goldman 2007; Hanchett 1997; Sachs 1992; Scott 1998; Sillitoe 2000) informs my analysis of government ideology and practice, the postcolonial state, and the foreign aid machinery that so dominates government practice in Bangladesh. The field of political ecology informs my analysis of environmental change in the region and the relationship of this change to the political structure, class relations, local knowledge and culture (Agarwal 1992; Blaikie 2012; Latour 2004; Mann 2009; Robbins 2012). I rely on human rights literature in my analysis of the hardships being faced by farming households in Chapra, and in developing recommendations for policy changes to mitigate these hardships (Irujo 2007; Khadka 2010; Klawitter and Qazzaz 2005; Langford 2005; O'Manique 1992; Pamukcu 2005; Sultana and Loftus 2012). Finally, I rely on water governance literature to determine the type of governance system best suited to protecting the human rights of all residents of the study area, while also sustaining the critical ecological characteristics of the river basin (Agrawal 1995; de Loë et al. 2009; Lebel et al. 2005; Orlove and Caton 2010; Wagner 2009, 2013).

Development theory

Development is a well-planned process for positive change from traditional to modern society for the goal of social, economic and political well-being (Gardner and Lewis 1996:3–5). This process is described as the top-down development approach that, by depriving marginalized communities, exploits institutional systems in formulating some specific programs and policies to gain greater control over local natural resources like fisheries, crops and water (Adger et al. 2005; Cleaver 1972:184; De Janvry 1975; Kottak 1999:29–30; Swain 1996; Van Ufford 1993:136–137). Gardner and Lewis (1996:7) argue that this management approach depicts unequal power relations between the North and the South. In this context, Nygren (1999:273) raises a major question: "Is development assistance a form of intellectual imperialism?"

In their historical review of development ideologies, Gardner and Lewis (1996) discuss the opposition that emerged during the 1960s between modernization and dependency theories. Contemporary debates differ from those of the 1960s in many ways, but carry forward some of the fundamental characteristics of that early debate. The modernization approach remains intact and continues to promote social transformation through industrialization and urbanization (Gardner and Lewis 1996:13). Level of industrialization is still used as one basic indicator of modernization following Rostow's model of economic growth. This approach continues to ignore the historical and political factors that cause underdevelopment (Gardner and Lewis 1996:15) and, arguably, the tendency of modernization approaches to cause underdevelopment has increased since the 1980s with the growing strength of neoliberal theory. The argument by dependency and world system theorists that colonialism laid the groundwork for underdevelopment in the global South (Amin 2003; Frank 1969; Prebisch 1963; Wallerstein 2004:x) remains relevant in the contemporary postcolonial era in which development agencies continue to depict the people of less developed nations as 'backward' and 'traditional.'

The continuing domination of the modernization paradigm with its perceived opposition of modernity and tradition creates epistemological differences that are often expressed today as an opposition between scientific knowledge and local knowledge. Many scholars have pointed out the ways in which this is a false dichotomy but they have also drawn attention to the fact that the disappearance of certain forms of local knowledge can serve the interests of powerful outside groups as well as local elites (Agrawal 1995; Brosius 1999; Latour 1987; Leach et al. 1999:225; Mosse 2013; Nygren 1999; Peet and Watts 1996b:11–16; Scott 1998:115). Another perceived form of opposition is that between top-down development approaches dependent on western scientific knowledge, imported technologies and development bank loans, and bottom-up development approaches using local knowledge. The most important message to be learned from an examination of these opposing discourses is that each works in support of or in opposition to a specific power structure (Gardner and Lewis 1996:25).

Goldman (2005) focuses on roles of the World Bank in shaping natural resource management all over the world. Based on this bank's approach, the linkage between community and nature is dominated by scientific knowledge that creates technological legitimacy to promote, for example, large watershed and Green Revolution based on the global Northern capital goods: e.g., dam, turbines, irrigation equipment, tractors, chemical fertilizers, pesticides and HYV seeds. Okongwu and Mencher (2000) argue that these technologies for resource management create ecocracies.

These ecocracies increased revenue collection of the developed countries (Gonigle 1999:16) whereas marginalized people of the peripheral countries encounter debt burdens that can be termed as "Aid as Imperialism, The Debt Trap" (Wood 1984:703) and "free-trade imperialism" (Makki 2004:151). If any third world country refuses to accept this development approach, the country could fail to receive foreign assistances (Skogly 1993:757), which could increase bankruptcy risk (Freytag and Pehnelt 2008:1). This development domination raises a major question: "development for whom?" (Nygren 1999:273) The major negative effects of this approach are a vicious cycle of underdevelopment and dependencies (Amin 2003; Frank 1969; Homer-Dixon 1994:391–400; Lewellen 1992; Ludden 2001:207; Martinez-Alier et al. 2010:153; Moore 2000; Prebisch 1963; Toye 1987:5; Wallerstein 2004). Based on this argument, I analyze here two major examples of Tanzania (Scott 1998) and Lesotho (Ferguson 1994).

Ferguson has provided an insightful analysis of how the discursive strategies of the Canadian International Development Agency (CIDA) and the World Bank (WB) informed the implementation of the Thaba-Tseka Development Project in 1975–84 in Lesotho. The project identified "backwardness," "lack of knowledge," "lack of credit" and the "traditional land tenure system" as key development limitations in this setting (Ferguson 1994:51). One description of this "backwardness" was that household was the basic unit of production and the only source of agricultural labor. Local exchanges of laborers, ploughs, cattle and seeds were also perceived as development limitations. The World Bank termed these limitations as transitions from traditional isolation to integration in international economy. Through the massive Thaba-Tseka Development Project, they sought to transform the region through intensification of cash crop production and integration of agricultural and non-agricultural development through reforms of local administrative, educational, economic and health institutions (Ferguson 1994:81); however, as Ferguson documented, these structural reforms increased inequality, decreased self-sufficiency, broke down local social and cultural rhythms of livelihoods and left people more vulnerable to 'natural' disasters, specifically drought.

Scott (1998) describes a similar development approach in Tanzania that he characterizes as following a western "high-modernist" agenda that reshapes the rules and regulations of statecraft, reconfiguring local practices to support this agenda. Following a utilitarian discourse and rational scientific forest management, state agencies reorganized the "seeding, planting, and cutting" of forest resources so that they were easier "to count, manipulate, measure and assess" (Scott 1998:13). This approach commercializes forest resources while ignoring locally embedded relationships between humans and forest that maintained complex, coordinated and sustainable ecological relationships.

Goldman (2007) demonstrates the ways in which the high modernist agenda of the World Bank is linked specifically to water management on a worldwide basis. During his five year (1968–73) tenure as the president of the World Bank, Robert McNamara financed more projects, 760, and loans, $13.4 billion,[1] than that of the combined investments of the Bank during all previous regimes. This investment provided essential support for the green revolution in the global South based on northern technologies, including irrigation technologies. Goldman (2007) describes this approach as green neoliberalism. The bank justifies its approach as providing support for food security and poverty reduction, while ignoring questions about its own self-interest and evidence about the negative outcomes of this top-down interventions with water commodification in project areas. Evidence of such negative effects is already well documented for Bangladesh, for instance. The World Bank supported programs like the Flood Action Plan (FAP) have created survival crises for the marginalized river bank in Bangladesh (Alexander et al. 1998; Banerjee 2010; Boyce 1990; Brammer 1990; Hanchett 1997; Mirza and Ericksen 1996; Paul 1999, 1995). Haque and Zaman (1993) have documented these negative effects for the Ganges Dependent Area (GDA) specifically. Hanchett et al. (1998:2) have documented the fact that rural women are the most vulnerable and bear the heaviest burdens in the case of these negative effects.

Within the field of development theory, the idea of ecocracy emerged during my study as a key concept that captures many of the arguments noted above. Sachs (1992) uses the term to describe the dominant approach of governments and international agencies such as the United Nations to the global crises of natural resource depletion and pollution. This approach is consistent with modernization theory, in that it relies on capital, bureaucracy and science to achieve better management and resolve these environmental crises. As problems escalate around the world, the "modernists" invent new technologies and programs with advanced models and integrated planning but with the ultimate goal of continuing business as usual. Escobar (1996) also uses the term ecocracy and, like Sachs, constructs it as part of his critique of the concept of sustainable development. Following the publication of *Our Common Future*, in 1987, the final report of the UN World Commission on Environment and Development, chaired by Gro Halem Brundtland, the term sustainable development came into prominence within the discourse of both environmentalism and development. But, according to Escobar and Sachs, the report promotes rapid economic growth and a technocentric vision of natural resource management. According to the report itself:

> The concept of sustainable development does imply limits, not absolute limits but limitations imposed by the present state of technology and social organization on environmental resources and by the ability of the biosphere to absorb the effects of human activities. But technology and social organization can be both managed and improved to make way for a new era of economic growth
>
> (World Commission on Environment and Development 1987:8).

The idea of ecocracy, as I use it in this book, draws on theory from all four of the theoretical fields I discuss in this chapter. Though developed initially as a critique of development, it bears heavily on the issue of governance and the relationship of human rights to ecological resource governance in particular.

While the above critiques of the dominant development paradigm provide insight into its many failures, they provide less insight into alternative approaches. Hanchett (1997) and Sillitoe (2000), however, do address the issue of alternative approaches and do so specifically in the context of studies they have conducted in Bangladesh. On the basis of her analysis of the negative outcomes of the Flood Action Plan, Hanchett (1997) argues that greater community participation in planning and implementation is necessary in order to avoid these negative outcomes. However, Hanchett argues, the current political and bureaucratic forms of controlled participation cannot change these outcomes. Despite flexibility of local knowledge, according to Dewan et al. (2014), "imposition of participation" cannot ensure proper representation and community empowerment with ecological resource management. In this context, Hanchett thinks that the World Bank and government should 'consult' with local people at the planning stages of a water development program. NGOs can be a helpful partner in this process.

Sillitoe (2000) emphasizes the importance of local and indigenous knowledge but is careful not to blame science and technology for the negative outcomes of the Flood Action Plan. He proposes that development agendas should be developed within communities. Community members are the experts when it comes to local ecological resources like soils, crops and water. Their knowledge is systematic and comprehensive and essential to local development. Based on close investigation in Bangladesh, Sillitoe (2000:4) argues that local knowledge is not stagnant and close-minded; rather it is "flexible, adaptable and innovative."

Political ecology

Several of the development theorists discussed above, Escobar in particular, are also well known for their contributions to political ecology. As Robbins (2012) points out, the field of political ecology has arisen from the confluence of several areas of study including cultural ecology, environmental history, political economy and development studies. Unlike development studies, however, political ecology necessarily involves an emphasis on the ecological as well as the political.

Piers Blaikie is a leading figure in the world of political ecology and is especially well known for his work on environmental degradation (Blaikie and Brookfield 1987). Of more immediate relevance to my book, however, is his more recent work on environmental justice movements. Blaikie notes the complex relationship among policy makers, politicians, consultancy firms, aid agencies, NGOs and civil society and the tendency for their collective efforts to exclude the voices of marginalized people from decision-making processes (Blaikie 2012). As a consequence, environmental justice movements have arisen, in the global south especially, and political ecologists have increasingly been drawn to document and support these movements. Blaikie cites the work of Robbins (1998) on the ecological aspects of development in Rajasthan, India, and his argument that engagement between southern and northern activists can produce a better outcome than purely local movements.

Some, though not all, political ecologists also closely examine the ways in which different social groups "construct" their ideas of "nature" and examine how those constructions support or undermine political structures and processes (Robbins 2012:122–138). Bruno Latour, however, in *Politics of Nature: How to Bring the Sciences into Democracy*, criticizes the field of political ecology for its tendency to promote an idea of nature that has been constructed through western science. Latour proposes the renewal of political ecology through the "depoliticization of science" (2004:7). This depoliticization will only be possible, he argues, when science comes out of modernism and negotiates the debate of hierarchy (Latour 2004:27–29).

Mann's approach to the construction of nature is consistent with Latour's but more productive for my book. Mann (2009) argues that political ecology should be more Gramscian in approach in order to incorporate the "ethico-political" element of Gramscian analysis. Ethico-politics in this context refers to the moral struggles that inform political debates. Mann (2009:336) thus argues that, "what is of interest is more than the discursive production of nature; it is a nothing less than a moral ecology." Mann proposes that the moral ecologies that oppose hegemonic processes become part of the historical process by which environmental change occurs and that political ecologists should align themselves with the moral ecologies that support environmental justice movements and oppose processes of marginalization. This approach is thus very consistent with that of Escobar where ecocracy is the most likely outcome the current hegemonic processes informing the sustainable development agenda.

My theoretical approach is also informed by the contributions of political ecologists to the analysis of environmental conflict and the protest movements that arise from environmental conflict. I did not originally intend to study protest movements at my fieldwork site but they were such a constant feature of life in Chapra that I could not ignore them. Protest movements arise in opposition to the hegemonic development processes I describe above but take many forms as accounted for in other settings by Escobar (1992), Kapoor (2011), Scott (1985) and Turner (1997). Scott's theories about "everyday forms of peasant resistance" and "weapons of the weak" (1985:49) are especially relevant to the pattern of resistance in Chapra. However, the pattern of resistance does not always correspond to what Scott has described for Malaysia. In many cases, people do organize themselves into formal movements, participating in street demonstrations, signing petitions and creating enduring organizations. Some of these movements in Bangladesh can be described in terms of Turner's (1997) parapolitical action, referring to actions "oriented above and beyond the political system of the state towards universal social, ethical and cultural values." Many of these movements also correspond to Escobar's (1992:27) description of "localized, pluralistic grassroots movements" (1992:27) that have many goals, attract people with diverse objectives and distrust conventional political parties and organizations. To understand these ecological movements, human rights approach to water governance can be helpful for protecting livelihoods of the marginalized people at Chapra.

Human rights

The Ganges River water right is the prerequisite for other human rights. From the UN (1948, 2002) human rights list, I base my research argument on five human rights components of food, work, education, health care and housing. Water right protection from the Ganges Basin flow can secure the ecosystem dependent agricultural production. Again, local people's rights to food (E/11), work (E/6), shelter (E/11), health care (E/12) and knowledge (E/13) described in the United Nations' Covenants can be protected with this eco-agricultural production (Khadka 2010:37).

Historically, the achievement of human rights is considered foundational to the progress of human civilization. The English Magna Carta (1215), the United States Declaration of Independence (1776), the French Declaration of the Rights of Man and the Citizen (1789), and the United Nations Universal Declaration of Human Rights (1948) are among the most important human rights documents (Khadka 2010). In addition to the Universal Declaration of Human Rights, the UN General Assembly has also passed declarations establishing the human right to water and also to food, work, health and housing. In line with other researchers who use UN definitions to evaluate the well-being of particular societies (Davidson-Harden et al. 2007; Khadka 2010:38; Langford 2005:275), my argument in this book is that access to water, specifically to Ganges River water right, is the prerequisite for other human rights.

Article 25 of the Universal Declaration of Rights focuses on standards of living and well-being and specifies the need for universal access to food, clothing, housing, health care and social services. The human right to water is first mentioned in the 1966 UN Covenant on Economic, Social and Cultural rights and in 2002 articles 11 and 12 of the Covenant were revised to include guidelines for protecting basic water rights (Dennis 2003; Irujo 2007; Khadka 2010; Klawitter and Qazzaz 2005:253; Langford 2005). International water law agreements such as the Helsinki Rules (1966), the UN Watercourses Convention (1997), and Berlin rules (2004) also lend support to the water as a human rights argument and provide detailed recommendations in support of the improved governance of international waters (Khadka 2010:42). Most recently, in 2010, the UN General Assembly adopted a further resolution recognizing "the right to safe and clean drinking water and sanitation as a human right that is essential for the full enjoyment of life and all other human rights" (General Assembly Resolution 64/292 as cited in Sultana and Loftus 2012:1).

The UN resolutions have limited legal authority even within signatory countries and, as Bakker (2012) and Sultana and Loftus (2012:10) emphasize, the right to water can mean very different things to different people. Nevertheless, a growing number of scholars are now evaluating the actions of governments and corporations in relation to the international standards outlined in international water agreements. Documentation of human rights, at a minimum, lends moral authority and political weight to the arguments of community organizations and social movements advocating for improved water access. Langford, for instance, notes that all households in South Africa have the right to receive 25 liters of free water, in accordance with the principle that water is a human right. Similarly, in Chile, following World Bank advice, water companies charge clients the full price for water but the government provides subsidies to water companies so they can give a basic amount of water free to low income families. Langford also notes, however, that implementation of these policies has been problematic in both countries (2005:277). In Ghana, by contrast, water prices doubled overnight, for rich and poor alike, when a public water utility was privatized, a move made by the national government in order to make the system attractive for foreign investors (Langford 2005:277).

Instead of focusing on this type of water privatization, Barlow and Clarke (2003) focus on the importance of water as a common property and raise concerns over human rights violations. Wetland is considered as the kidney of freshwater ecosystem and can be protected only with this common property approach (Barlow 2008:21). On the other hand, the root cause of ecological crises is river flow degradation that is termed as "water apartheid" in the global South (Barlow 2008:5).

Irujo (2007) argues that water rights protection is a major prerequisite for other human rights based on the United Nations (2002) principles of respect, protect and fulfill. Klawitter and Qazzaz (2005:253) focus on the status of human rights in several Middle Eastern countries emphasizing on water quantity, quality and price for domestic use. They argue that the right to water should be supported because of its importance to human survival, security and capability construction (Klawitter and Qazzaz 2005:255). Water management should be set up, they argue, so as to ensure these freedoms based on an equality and social justice approach. From this perspective governments are also responsible for *protecting* their citizens from the influence of a third party like the International Monetary Fund or the World Bank. Neoliberal economic policy and structural adjustment programs should not be allowed to diminish the right to water.

My approach in this book extends the arguments discussed above to emphasize the government's obligation to care for and preserve the ecosystems on which communities like Chapra depend. As part of this obligation, the government needs to be conscious about negative effects of the top-down domination over the understanding of human rights, and therefore, Chase (2016) emphasizes on bottom-up approach of this understanding. Also, as Pamukcu (2005) argues, the government obligation to protect water access and all the rights that depend on it, extends to international rivers and implies the need for cooperation among all countries sharing a river basin.

Pamukcu (2005) argues that water as a human right should be affordable, easily accessible and safer for personal and domestic usage. The process of securing this water right is an obligation to every national government in the context of an international river basin. In the context of International River, the obligation implies to every basin country and to the international community (Pamukcu 2005:160). Pamukcu (2005) also focuses on vital human needs of the 1997 United Nations Watercourses Convention as an important aspect of water as a human right. The basin ecosystem protection can promote this water right that can protect other human rights (Pamukcu 2005) based on UN (2002) principles of human rights. In this context, the Ganges Basin water has the power to boost up these human rights and community livelihoods and to enable a sustainable future for the basin countries. Rasul (2014a) describes the Hindu Kush Himalayan ecosystem services in protecting the downstream regions' food, water and energy security. To ensure this future security, Sultana and Loftus (2012) focus on water governance for the goal of water rights protection based on principles of water justice and equity.

Water governance

Water governance approach can be described based on the ecological perspectives of Baviskar (2007) and Wagner (2013). Baviskar (2007) uses the concept of *waterscape* to describe power relations in water resource management. Water as a flowing resource and land as a stable resource shape local culture, community and livelihood. This ecosystem governance needs to be promoted against the technocentrisms. Corporate elites promote this technocentric water development that not only changes waterscape but also disturbs harmony between community and nature. This domination over nature, according to Agarwal (1992:120), can be overcome by recognizing locally embedded understanding of ecological resource management. Using Marxism and Gandhian approach, Agarwal (1992:145–146) wants to secure environmental protection and local development based on the marginalized community participation (Agarwal 1992:145–146).

The recognition of community participation in water policy formulation and implementation process is the most important points of water governance. This governance approach requires incorporation of local ecosystem. Wagner (2013:7) based on the notion of "human-water relations" analyzes "socioecological systems":

> if our goal is to live sustainably within the watersheds we inhabit, then to understand watersheds as socioecological systems, as whole systems, not systems that are sometimes social and sometimes ecological, and not always both at the same time.

This socioecological system offers a specific form of community livelihoods (Abel and Stepp 2003; Brosius 1999). Hastrup (2013), emphasizing on ecosystem, argues that water forms a particular societal world based on connection among river flows, seasonal patterns, cropping practices, vegetation and habitat. I describe these linkages in the context of the Ganges Basin ecosystems that can protect local cultural practices about agricultural production, employment practices and food consumption.

The *borsha* flow provides major ecological services such as cropland siltation at Chapra which are similar to the concept of common pool resources (Ostrom et al. 1999). Paul (1984:15) describes this *borsha* flow as an important part of producing some crops like rice paddy. Again, the *khora* season helps to produce the different local crops, wild vegetables and fertilizers. This ecosystem practice follows the theoretical analyses of Abel and Stepp (2003), Clay and Lewis (1990), Johnston (2010) and Vayda and McCay (1975). According to Clay and Lewis (1990), local ecosystem based agricultural practices are helpful for erosion control and for reducing land degradation in Rwanda. Chapra villagers follow this ecological conservation and livelihood practices based on seasonal cropping patterns (Craig et al. 2004; Cuny 1991).

This ecosystem approach can neutralize the current hydropolitics in the Ganges Basin (Brichieri-Colombi and Bradnock 2003). This approach considers the basin as a unit to harmonize the relationships between water, land and community (Abel and Stepp 2003; Brunnee and Toope 1997) based on ecological context of seasonal riverflow (Chowdhury and Ward 2004), floodplain fisheries (Craig et al. 2004), vegetation (Crawford 2003) and watershed (Isaak and Huber 2001). This harmony can protect ecosystem of the Himalaya Mountains based on ecological integrity of the Ganges-Brahmaputra Basin flow in China, India, Nepal and Bangladesh (Pasi and Smardon 2012). This integrity can be described as waterscapes ecology that is connected with landscapes and political ecology based on watershed as a unit that ranges from local community to national government and international institutions (Orlove and Caton 2010).

This integrity protection requires water governance based on community participation (Dove 1986) in the context of ecosystem (Folke et al. 2005; Hukka et al. 2010; Mirumachi and Allan 2007; Wagner 2010) and environmental governance (Bulkeley 2005) that can be the major foundation for common pool resource management (Bardhan 1993; Castro 2007; Ostrom et al. 1999) and for restoring resource conservation (Agrawal and Gibson 2001). These ecological and livelihood linkages confirm that water is the total social fact and can be promoted with water governance.

Orlove and Caton's (2010) water governance approach describes water as "total social facts" that signify social, economic, political and cultural domains of life based on social organizations that range from local community to international institutions. These domains connect natural resources and human rights. One major focusing point of this connection is water and agricultural production linkages based on the concept of political economy. The direction of this connection depends on the roles of political governments which can follow top-down or bottom-up approach.

Sillitoe (1998:204) defines local knowledge as socio-cultural traditions for promoting the bottom-up approach which are ascribed by birth and are helpful for coping with local environments. I use this definition to explain local farming communities or *chashi* (traditional farmers) who describe *borsha* (wet) and *khora* (dry) seasons as normal ecological processes every year which maintain local agro-ecological and employment practices. *Chashi* transmit this knowledge from one generation to another to adapt with local culture and environment. This transmission system is occurred with locally developed social organizations (Bebbington 1993). Mosse (1997:1) defines this organization based on Bourdieu's concept of "symbolic capital" that articulates, reproduces and challenges social relations. The Ganges Basin ecosystems are the prerequisite for this capital that is articulated through common pool ecological resources. This articulation develops community rules, norms and conventions (Biersack 1999:9; Harris 1966; Kottak 1999:23). These conventions need to be recognized in water governances at the basin, national and local levels based on water democracy (Shiva 2005). For this democratic development, science needs to come out of corporate control (Latour 2004) so that local knowledge can get recognition as an important part of the scientific world. Here, environmental governance theory broadens the scopes of how resources could be managed to include much more than the government regulations created as part of a formal management system. This broader framework includes the roles of community members, advisory groups, interest groups, lobbyists and NGOs (de Loë et al. 2009; Wagner 2009). Governance theory also recognizes the role of values in shaping decision-making processes and the fact that different governance approaches articulate with different types of human/water relationships. Hastrup (2013), similarly, argues that "waterworlds," come into being based on the connections among river flows, seasonal patterns, vegetation and habitat and cultural practices.

A socioecological approach to governance is relevant to my research because it could help promote community voices in water management and support an ecosystem-based approach. The ecosystem concept has its limitations, as many scholars have pointed out, but it is appropriate for my research because of the deep dependence of marginalized households in Chapra on the ecosystem services of the Ganges and Gorai River basins. An ecosystem approach that is inclusive of the social relations embedded within it, is consistent with the observation by Orlove and Caton (2010:402) that water is a "total social fact" that needs to be understood holistically. On that basis they critically analyze the Integrated Water Resources Management (IWRM) approach because of its tendency to integrate only specific resource sectors in management processes and to limit IWRM boundaries to national boundaries (Orlove and Caton 2010). Criticizing overly technocentric and bureaucratic approaches, they argue for inclusion of cultural factors and the integration of different sectors and groups at regional, national and international levels. Rather than the World Bank's evaluation of water as a commodity, Orlove and Caton evaluate water as a human right and emphasize the social justice aspect of water policy and governance.

Wagner (2009) recommends a distributive, multi-level water governance approach that facilitates the involvement of both formal and informal institutions in decision-making processes. He argues that this approach will be more effective than centralized, state-dominated water management systems if the goal is to preserve ecological resilience as well as economic sustainability. Distributing authority among institutions at different levels within the total system, he argues, is also more democratic and can create a system of checks and balance among competing groups. Based on his research in the Okanagan Valley in British Columbia, Wagner notes that governance systems, which are highly centralized in formal terms, can behave more like distributed systems in practice, but that legislative changes should be made to acknowledge the principle of joint authority by institutions at different levels and to discourage unilateral decision-making by central authorities (2009:5).

Lebel et al. (2005) focus more explicitly than Wagner, on the question of scale in their study of Mekong Delta water governance, noting the simultaneous presence of spatial, temporal and jurisdictional scales. Multi-scalar institutional networks generate their own unique political dynamics and different actors' interests "constrain, create and shift scales and levels" (Lebel et al. 2005:2). Governance networks are strongly influenced by local development history, by international development institutions like the World Bank, class relations and by hydrological and ecological factors including climate change. Ideally, a multi-scalar system will include mechanisms for including marginalized groups at different scales and will provide them with real power in decision-making processes (Rasul 2014b). However, this multi-scaler system has formal and informal dynamics that creates difficulties in sustainable water resource management for agricultural production (Afroz et al. 2016). Dewan et al. (2014) explain these difficulties with the concept, "imposition of participation," that is not suitable for community representation and empowerment when local knowledge describes many things to the different people. To address these concerns properly, the governance approaches that are most informative for my research are those that focus on social justice and human rights issues and on the principles of equity and environmental sustainability, as opposed to the more narrow approaches that focus on water as an economic resource alone. To understand this socioecological and economic dynamics, an extensive ethnographic fieldwork was done at Chapra, Kushtia District in Bangladesh.

Chapra village as the fieldwork site

Chapra was selected as the fieldwork site because of its location on the Gorai River and because it has experienced significant environmental impacts during both wet and dry seasons due to the Farakka diversion. Chapra is only five kilometers away from the Ganges River (Map 1.2).

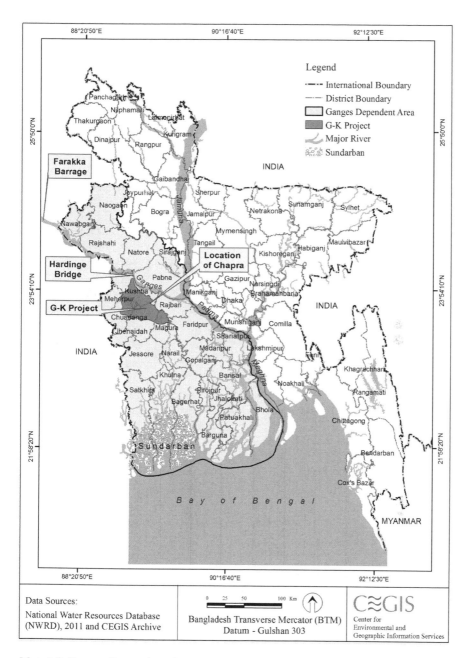

Map 1.2 Ganges Dependent Area and fieldwork site at Chapra

The area is also subject to the influences of two large-scale water engineering projects implemented by the government of Bangladesh which are currently used to mitigate the effects of the Farakka diversion. Farming households at Chapra are currently dependent for much of their irrigation water on the Ganges-Kobodak (GK) Project, the largest irrigation project in Bangladesh occupying an area of about 1700 square kilometers in four districts (Bangladesh Water Development Board n.d.). In some locations the project canals are located only five hundred feet away from the site of another major project, the Gorai River Restoration Project (Map 1.2). Chapra is thus subject to multiple large-scale water engineering projects that exemplify the scientific and technology-driven approach to both the Indian and Bangladeshi governments; and the impacts of these projects on hydro-ecology, community livelihoods and human rights is broadly representative of what is happening throughout the region.

Chapra is a rural village that is composed of six *paras* or sub-villages: Bohla Govindpur, Dekipara, Koburat, Madulia, Charpara and Purbopara, and Sindoh. With a population of 4,331 people, Chapra is the largest of several villages under the jurisdiction of Chapra Union Council and it falls within the Kumarkhali Upazila Parishad (Sub-district) within Kushtia District. Chapra village is about 236 kilometers away from the Farakka Barrage in India and about 11 kilometers downstream from the Hardinge Bridge in Kushtia where the government in Bangladesh has established one of its data collection stations of the Ganges flow. The Ganges River enters into Bangladesh in Rajshahi District and passes through Kushtia District before it merges with the Brahmaputra River at Shibalya Upazila in Manikgonj District.

About 70 percent of the population of Chapra (963 of 1387 households) make their living through farming or farm labor (Bangladesh Election Commission 2011). The Bangladesh Bureau of Statistics (2005) classifies farming households as large, medium, small and marginalized. Based on this classification, I describe in this book the large and medium farmers as the rich and intermediate farmers respectively. Rich farmers are those who own 7.50 acres or more agricultural land; intermediate farmers own between 2.50 and 7.49 acres; small farmers own between 0.50 and 2.49 acres; marginalized farmers own less than 0.49 acres of cropland. Rich farmers make up one percent of the population; intermediate farmers make up eight percent; small farmers 51 percent and marginalized farmers 41 percent (Bangladesh Bureau of Statistics 2005). Fourteen percent of the marginalized farmers own no farm land at all but depend on farm labor for all or most of their income. In this context, this study refers to both small and marginalized farmers as marginalized throughout this book since 2.49 acres of land are not enough for a household's self-sufficiency.

Primary data collection

Focus groups

As a major data source, the four focus groups were convened, each stratified by socioeconomic position, to gather qualitative data about local water practices, seasonal patterns, basin history and community livelihoods at Chapra. Many people have lived in Chapra for seven generations or more and the focus groups allowed me to gather information about these generational water practices and present challenges. The target groups were local farmers, of varied socio-economic status, and agricultural laborers whose livelihoods are dependent on the basin ecosystems (Adnan 1991; Boyce 1990; Wood 1999:731). A significant number of women attended the focus group with their male household members. This study includes individuals over 40 years of age since they are more likely to have in-depth knowledge about past water practices, ecosystems and local livelihoods. Children below 18 years of age were excluded from the focus group. Local research assistants helped me to arrange the focus group meetings.

Forty-four male and female household heads attended the focus group meeting, including 7 rich farmers, 9 intermediate farmers and 28 marginalized farmers, 12 of whom were small farmers and 16 day laborers who represent proportionately the rich, intermediate and marginalized farmers. One research assistant was assigned to the group of 7 rich farmers, 1 to the 9 intermediate farmers, and 2 assistants to each of the larger groups of 12 small farmers and 16 day laborers. Local dialects were used as the main means of communication with the focus participants. With their verbal permission, their conversations were recorded and pictures were taken. At the end of focus group meeting, the four community members from the different socioeconomic conditions were selected for in-depth case studies from 2011 to 2012.

In-depth case studies and participant observation

In order to gain as much insight as possible into traditional agricultural practices, farming and environmental history, I selected the four older farmers and laborers who possessed leadership qualities and extensive local knowledge. Based on these criteria, I selected Tanvir, Tofajjal, Joardar and Billal, who represented the rich, intermediate, marginalized and traditional day laborer socioeconomic groups, respectively. I used these pseudonyms to protect their privacy.

Tanvir

Tanvir is a 60-year-old rich farmer, who has lived in his entire life at Chapra. He completed the Higher Secondary Certificate (HSC) educational degree but did not apply for an office job because he received 23 acres of cropland from his parents and another 7 acres from his wife's parents. He married Sohali in 1970. Tanvir and Sohali have three sons, Kofil, Masud and Firoj, and two daughters, Shapla and Shaon. Currently, Tanvir makes profits on croplands, fruit gardens, fisheries, commercial agricultural production and agribusiness. He has five fisheries projects and four fruit gardens. He uses leased land from the government sponsored Ganges-Kobodak project for two of his fisheries projects and low cropland of his own for the other three. Most of his assets are within four kilometers of his house, the Gorai River and the Ganges-Kobodak project. He also has two irrigation machines, two plow machines, two mechanized vans, eight mechanized paddy machines and three crop storage facilities. Moreover, he has three residences, one each in Chapra, the nearby city of Kushtia and the capital city, Dhaka. His houses in Kushtia and Dhaka are four- and three-floor buildings respectively.

Tofajjal

Tofajjal is a 59-year-old intermediate farmer who has lived in Chapra since his birth. In addition to his current house, he has a second house in Chapra and has built a three-floor apartment building in Kushtia. He has completed a School Secondary Certificate degree but his parents did not allow him to apply for an office job. He received four acres of cropland from his parents. He married Salma in 1979 and obtained another three acres from his father-in-law. Tofajjal and Salma have two daughters, Shahida and Karima, and two sons, Mamun and Kabir. Tofajjal has two fish ponds and three fruit gardens. He has a shallow tube-well, one plow machine and two mechanized paddy machines. Currently, Tofajjal is successful in maintaining a decent livelihood based on his own resources.

Joardar

Sixty-five-year-old Joardar is a marginalized farmer whose family has lived in Chapra for more than six generations. He dropped out from school in the ninth grade and now owns half-an-acre cropland. His cropland is one kilometer away from his house and the Gorai River. He married Suma Khatun in 1971, and his wife's family has also lived in Chapra for more than four generations.

Billal

Billal is a 52 year old, traditional agricultural day laborer whose family has lived in Chapra for three generations. He has neither a formal school education nor a piece of cropland. Agricultural day labor is his main source of income. His house is 20 meters away from the Gorai River. His wife, Parvin, has also lived in Chapra since her birth but her father is originally from another district. They have four daughters, Jesmin, Kamrun, Shampa and Shakina, and three sons, Jamal, Lablu and Kamal. Jesmin is 23 years old and does not have formal education. She is married to Kubbat, who works as a van driver. Kamrun, 25 years old, dropped out from school after the sixth grade and is married to Khokon, who works as an agricultural worker. Shampa, age 23, dropped out of school after the fifth grade and is married to Sabbir, who works as an agricultural worker and fisherman. Jamal, age 21, has migrated to Dhaka, where he works as a garment worker. Kamal, 18 years of age, works for Tanvir.

For data collection from these four people, my male research assistants and I were closely attached to the male case respondents and my female research assistants were closely attached to the female household members. I kept records of all activities and took photographs of the basin flow, the Gorai River's shallow condition, river bank erosion, embankment failures, fisheries projects, agricultural production and local water bodies.

Chapra household survey

The household survey script was comprised of 185 questions developed based on experiences of focus group discussions in order to collect quantitative data on current water practices, agricultural challenges and human rights conditions at Chapra. Bangladesh Election Commission voter list for Chapra was helpful to select the household survey respondents. From this list, 259 of 963 agricultural households were selected based on simple random sampling using a lottery system without replacement (26.9 percent of all agricultural households). Among these households, there were 34 women respondents who originated from the marginalized households. Seven rich (2.7 percent of my sample but 100 percent of the rich farmers in Chapra) and 14 intermediate farming households (5.4 percent of my sample but 22.4 percent of all intermediate farmers in Chapra) for this survey.

Secondary data sources

In addition to the primary data sources, an extensive body of secondary data about water management practices, policies and projects was gathered from government records and publications at local, regional and national levels. At the national level, I collected secondary data directly from several government ministries including the Ministry of Water Resources (MoWR), Bangladesh Water Development Board (BWDB), Department of Agricultural Extension (DAE), Ministry of Agriculture (MoA), Bangladesh Bureau of Statistics (BBS), Water Resources Planning Organization (WARPO), and Ministry of Disaster Management and Relief (MoDMR). For example, the BWDB provided me the Ganges River flow data at the Hardinge Bridge Station for 1960–2011 and detailed documentation concerning the Gorai River Restoration Project and the Ganges-Kobodak project.

Conclusion

The secondary data along with fieldwork evidence were helpful for describing the ecological characteristics and hydropolitics. Given the large area of the Farakka impacts within Bangladesh, one that extends from the border of India and Bangladesh to the point where the now much diminished Ganges River joins the larger flow of Brahmaputra, it was necessary to select a representative but delimited area of my fieldwork. For this reason, this research project focused on Chapra, a village in Kushtia District, as suitable for the research goals. The diversion of water in India by the Farakka Barrage is the single most immediate cause of water shortages and associated hardships throughout the Ganges Dependent Area in Bangladesh but my study indicates that the developmentalist ideologies of both India and Bangladesh are the root causes of socioecological concerns over Chapra. Their water management approaches are top-down, neoliberal, technocratic and ecocratic and favor elite interests over the interests of the majority. These approaches decrease the quantity of ecological resource commons and, through privatization and commoditization, reduce the access of marginalized groups to the traditional resources that remain and to the new resources created by new technologies. This common versus privatization is described with theoretical approaches of development theory, political ecology, human rights and water governance. From these perspectives the Farakka Barrage diversion is best understood as just one of many expressions of the underlying cause with the GK Project and GRRP providing two additional examples. Although my book focuses on the local effects of government interventions at Chapra, the scale of the interventions has necessarily required some consideration of more distant impacts. Based on this argument, it is important to focus on the Ganges Dependent Area (GDA) in Bangladesh to describe deeper effects of these interventions.

Note

1   All dollar figures in this book are given in US dollars unless otherwise stated.

References

Abel, T. and J. R. Stepp 2003. Editorial: A New Ecosystems Ecology for Anthropology. *Conservation Ecology* 7(3): 12.

Adger, W. N., K. Brown and E. L. Tompkins 2005. The Political Economy of Cross-Scale Networks in Resource Co-Management. *Ecology and Society* 10(2): 9.

Adnan, S. 1991. *Floods, People and the Environment.* Dhaka: Research and Advisory Services.

Afroz, S. 2012. Akij Bidi Factory Unrest: 2 Workers Shot Dead in Kushtia. Dhaka: *Bdnews.* July 24, 2015. Accessed May 9, 2013

Afroz, S., R. Cramb and C. Grunbuhel 2016. Collective Management of Water Resources in Coastal Bangladesh: Formal and Substantive Approaches. *Human Ecology* 44(1): 17–31.

Agarwal, B. 1992. The Gender and Environment Debate: Lessons from India. *Feminist Studies* 18(1): 119–158.

Agrawal, A. 1995. Dismantling the Divide between Indigenous and Scientific Knowledge. *Development and Change* 26(3): 413–439.

Agrawal, A. and C. Gibson 2001. *Communities and the Environment: Ethnicity, Gender, and the State in Community-based Conservations.* New Brunswick: Rutgers University Press.

Alexander, M. J., M. S. Rashid, S. D. Shamsuddin and M. S. Alam 1998. Flood Control, Drainage and Irrigation Projects in Bangladesh and Their Impacts on Soils: An Empirical Study. *Land Degradation & Development* 9(3): 233–246.

Altinbilek, D. 2002. The Role of Dams in Development. *Water Resources Development* 18(1): 9–24.

Amin, S. 1974. *Accumulation on a World Scale: A Critique of the Theory of Under-development*. Volume 2. New York and London: Monthly Review Press.

—— 2003. *Obsolescent Capitalism: Contemporary Politics and Global Disorder*. London: Zed Books.

Bakker, K. 2012. Commons versus Commodities: Debating the Human Right to Water. In *The Right to Water: Politics, Governance and Social Struggles*. F. Sultana and A. Loftus, eds. Pp. 19–44. London and New York: Earthscan.

Bandyopadhyay, J. 2006. Integrated Water Systems Management in South Asia: A Framework for Research. *CEP Occasional Paper 09*. Calcutta: Indian Institute of Management.

Bandyopadhyay, J. and N. Ghosh 2009. Holistic Engineering and Hydro-Diplomacy in the Ganges-Brahmaputra-Meghna Basin. *Economic and Political Weekly* 44(45): 50–60.

Banerjee, L. 2010. Creative Destruction: Analyzing Flood and Flood Control in Bangladesh. *Environmental Hazards* 9(1): 102–117.

Bangladesh Bureau of Statistics (BBS) 2005. *Key Information of Agriculture Sample Survey*. Dhaka: Government of Bangladesh.

—— 2010a. *Area Population Household and Household Characteristics*. Dhaka: Government of Bangladesh.

—— 2010b. *Household Income and Expenditure Survey (HIES)*. Dhaka: Government of Bangladesh.

Bangladesh Election Commission (BEC) 2011. *Voter List of Chapra Union at Kushtia District*. Dhaka: Government of Bangladesh.

Bangladesh Water Development Board (BWDB) N.d. Ganges-Kobodak (GK) Irrigation Project. Dhaka: Government of Bangladesh.

Bangladpedia 2012. Ganges-Kobodak Irrigation Project. http://www.banglapedia.org/HT/G_0029.htm, accessed July 3, 2013.

Bardhan, P. 1993. Symposium on Management of Local Commons. *Journal of Economic Perspectives* 7(4): 87–92.

Barlow, M. 2008. *Blue Covenant: The Global Water Crisis and the Coming Battle for the Right to Water*. New York: New Press.

Barlow, M. and T. Clarke 2003. *Blue Gold: The Battle against Corporate Theft of the World's Water*. London: Earthscan.

Baviskar, A. ed. 2007. *Waterscapes: The Cultural Politics of Natural Resource*. Delhi: Permanent Black.

Bebbington, A. 1993. Modernization from Below: An Alternative Indigenous Development? *Economic Geography* 69(3): 274–292.

Bharati, L. and P. Jayakody 2011. Hydrological Impacts of Inflow and Land-use Changes in the Gorai River Catchment, Bangladesh. *Water International* 36(3): 357–369.

Bhattarai, D. P. 2009. An Analysis of Transboundary Water Resources: A Case Study of River Brahmaputra. *Journal of the Institute of Engineering* 7(1): 1–7.

Biersack, A. 1999. The Mount Kare Python and His Gold: Totemism and Ecology in the Papua New Guinea Highlands. *American Anthropologist* 101(1): 68–87.

Biswas, A. K. and C. Tortajada 2001. Development and Large Dams: A Global Perspective. *Water Resource Development* 17(1): 9–21.

Blaikie, P. 2012. Should Some Political Ecology Be Useful? The Inaugural Lecture for the Cultural and Political Ecology Specialty Group, Annual Meeting of the Association of American Geographers, April 2010. *Geoforum* 43(2): 231–239.

Blaikie, P. and H. Brookfield 1987. *Land Degradation and Society*. London: Methanuem.

Blair, H. W. 1978. Rural Development, Class Structure and Bureaucracy in Bangladesh. *World Development* 6(1): 65–82.

Boyce, J. K. 1990. Birth of a Mega Project: Political Economy of Flood Control in Bangladesh. *Environmental Management* 14(4): 419–428.

Brammer, H. 1990. Floods in Bangladesh: II. Flood Mitigation and Environmental Aspects. *The Geography Journal* 156(2): 158–165.

Brichieri-Colombi, S. 2008. Could Bangladesh Benefit from the River Linking Project? In *Interlinking of Rivers in India: Issues and Concerns*. Mirza, M.M.Q., A. U. Ahmed and Q.K. Ahmed. Pp 275-289. Leiden: CRC Press/Balkema.

Brichieri-Colombi, S. and R. W. Bradnock 2003. Geopolitics, Water and Development in South Asia: Cooperative Development in the Ganges-Brahmaputra Delta. *Royal Geographic Society with IBG* 169(1): 43–64.

Brosius, J. P. 1999. Analyses and Interventions: Anthropological Engagements with Environmentalism. *Current Anthropology* 40(3): 277–310.

Brunnee, J. and S. J. Toope 1997. Environmental Security and Freshwater Resources: Ecosystem Regime Building. *The American Journal of International Law* 91(1): 26–59.

Bulkeley, H. 2005. Reconfiguring Environmental Governance: Towards a Politics of Scales and Networks. *Political Geography* 24: 875–902.

Castro, J. E. 2007. Water Governance in the Twentieth-first Century. *Campinas* X(2): 97–118.

Chase, A. T. 2016. Human Rights Contestations: Sexual Orientation and Gender Identity. *The International Journal of Human Rights*. DOI: 10.1080/13642987. 2016.1147432.

Chowdhury, M. R. and N. Ward 2004. Hydro-Meteorological Variability in the Greater Ganges-Brahmaputra-Meghna Basins. *International Journal of Climatology* 24: 1495–1508.

Clay, D. C. and L. A. Lewis 1990. Land Use, Soil Loss, and Sustainable Agriculture in Rwanda. *Human Ecology* 18(2): 147–161.

Cleaver, H. M. 1972. The Contradictions of Green Revolution. *The American Economic Review* 62(1&2): 177–186.

Cox, R. 1987. *Power, Production, and World Order: Social Forces in the Making of History*. New York: Columbia University Press.

Craig, J. F., A. S. Halls, J. J. Barr and C. W. Bean 2004. The Bangladesh Floodplain Fisheries. *Fisheries Research* 66: 271–286.

Crawford, R. C. 2003. Riparian Vegetation Classification of the Columbia Basin, Washington. *Natural Heritage Program Report 2003–03*. Olympia, WA: Washington Dept. Natural Resources.

Crow, B. and N. Singh 2000. Impediments and Innovations in International Rivers: The Waters of South Asia. *World Development* 28(11): 1907–1925.

Cuny, F. C. 1991. Living with Floods-Alternatives for Riverine Flood Mitigation. *Land Use Policy* 8(4): 331–342.

Das, V. and D. Poole eds. 2004. *Anthropology in the Margins of the State*. Santa Fe: School of American Research Press.

Davidson-Harden, A., A. Naidoo and A. Harden 2007. The Geopolitics of the Water Justice Movement. *Peace Conflict & Development* 11: 1–34.

De Janvry, A. 1975. The Political Economy of Rural Development in Latin America: An Interpretation. *American Journal of Agricultural Economies* 57(3): 490–499.

deLoë, R. C., D. Armitage, R. Plummer, S. Davidson and L. Moraru 2009. *From Government to Governance: A State-of-the-art Review of Environmental Governance.* Guelph: Rob de Loë Consulting Services. Prepared for Alberta Environment, Environmental Stewardship, Environmental Relations.

Dennis, M. J. 2003. Human Rights in 2002: The Annual Sessions of the UN Commission on Human Rights and the Economic and Social Council. *The American Journal of International Law* 97(2): 364–386.

Dewan, C., M. C. Buisson and A. Mukherji. 2014. The Imposition of Participation? The Case of Participatory Water Management in Coastal Bangladesh. *Water Alternatives* 7(2): 342–366.

Dove, M. R. 1986. Peasant Versus Government Perception and Use of the Environment: A Case-Study of Banjarese Ecology and River Basin Development in South Kalimantan. *Journal of Southeast Asian Studies* 17(1): 113–136.

D'Souza, R. 2006. Water in British India: The Making of a 'Colonial Hydrology'. *History Compass* 4(4): 621–628.

Elhance, A. P. 1999. *Hydro-politics in the Third World: Conflict and Cooperation in International River Basins.* Washington: United States Institutes of Peace Press.

Escobar, A. 1991. Anthropology and the Development Encounter: The Making and Marketing of Development Anthropology. *American Ethnologist* 18(4): 658–682.

——— 1992. Imagining a Post-Development Era? Critical Thought, Development and Social Movements. *Social Text* 31/32: 20–56.

——— 1996. Construction Nature: Elements for a Post-structuralist Political Ecology. *Futures* 28(4): 325–343.

Ferguson, J. 1994. *The Anti-Politics Machine: Development, Depoliticization, and Bureaucratic Power in Lesotho.* Minneapolis: University of Minnesota Press.

Folke, C., T. Hahn, P. Olsson and J. Norberg 2005. Adaptive Governance of Socio-Ecological Systems. *Annual Review of Resource* 30: 441–473.

Frank, A. G. 1969. *Capitalism and Underdevelopment in Latin America: Historical Studies of Chile and Brazil.* New York: Monthly Review Press.

Freytag, A. and G. Pehnelt 2008. *After the Crisis Is Before the Crisis: The Political Economy of Debt Relief.* Working Papers on Global Financial Markets No. 2.

Gardner, K. and D. Lewis 1996. *Anthropology, Development and the Post-modern Challenge.* London: Pluto Press.

Gilmartin, D. 1994. Scientific Empire and Imperial Science: Colonialism and Irrigation Technology in the Indus Basin. *The Journal of Asian Studies* 53(4): 1127–1148.

Giordano, M., M. Giordano and A. Wolf 2002. The Geography of Water Conflict and Cooperation: Internal Pressures and International Manifestations. *The Geographical Journal* 168(4): 293–312.

Gleick, P. H. 2009. Facing Down the Hydro-Crisis. *World Policy Institute* 26(4): 17–25.

Goldman, M. 2005. *Imperial Nature: The World Bank and Struggles for Social Justice in the Age of Globalization.* New Haven, CT: Yale University Press.

Goldman, M. 2007. How "Water for All" Policy Became Hegemonic: The Power of the World Bank and Its Transnational Policy Networks. *Geoforum* 38: 786–800.

Gupta, A. D., M. S. Babel, X. Albert and O. Mark 2005. Water Sector of Bangladesh in the Context of Integrated Water Resources Management: A Review. *International Journal of Water Resources Development* 21(2): 385–398.

Gyawali, D. and A. Dixit 2001. Water and Science: Hydrological Uncertainties, Developmental Aspirations and Uningrained Scientific Culture. *Futures* 33(8&9): 689–708.

Haftendorn, H. 2000. Water and International Conflict. *Third World Quarterly* 21(1): 51–68.

Hanchett, S. 1997. Participation and Policy Development: The Case of the Bangladesh Flood Action Plan. *Development Policy Review* 15(3): 277–295.

Hanchett, S., J. Akhter and K. R. Akhter 1998. Gender and Society in Bangladesh's Flood Action Plan. In *Water, Culture, and Power: Local Struggles in a Global Context*. J. M. Donahue and B. R. Johnston eds. Pp. 209–234. Washington, DC: Island Press.

Haque, C. E. and M. Q. Zaman 1993. Human Responses to Riverine Hazards in Bangladesh: A Proposal for Sustainable Floodplain Development. *World Development* 21(1): 93–107.

Harris, M., N.K. Bose, M. Klass, J. P. Mencher, K. Oberg, M. K. Olper, W. Suttles, and A. P. Vayda 1966. The Cultural Ecology of India's Sacred Cattle [and Comments and Replies]. *Current Anthropology*, 7(1): 51–66.

Hastrup, K. 2013. Water and the Configuration of Social Worlds: An Anthropological Perspective. *Journal of Water Resource and Protection* 5: 59–66.

Hill, D. 2006. The Politics of Water in South Asia. *Transforming Cultures eJournal* 1(2): 136–158.

Homer-Dixon, T. F. 1994. Environmental Scarcities and Violent Conflict: Evidence from Cases. *International Security* 19(1): 5–40.

Hossen, M. H. 2012. Bilateral Hydro-hegemony in the Ganges-Brahmaputra Basin. *Oriental Geographer* 3(53): 1–18.

——— 2017 (forthcoming). Ecological Integrity of the Brahmaputra Basin for Community Livelihoods in Bangladesh. *Journal of the Asiatic Society of Bangladesh*.

Hukka, J. J., J. E. Castro and P. E. Pietila 2010. Water, Policy and Governance. *Environment and History* 16: 235–251.

Irujo, A. E. 2007. The Right to Water. *Water Resources Development* 23(2): 267–283.

Isaak, D. J. and W. A. Hubert 2001. Production of Stream Habitat Gradients by Montane Watersheds: Hypothesis Tests Based on Spatially Explicit Path Analyses. *Canadian Journal of Fisheries and Aquatic Sciences* 58: 1089–1103.

Islam, G. M. and M. R. Karim 2005. Predicting Downstream Hydraulic Geometry of the Gorai River. *Journal of Civil Engineering (IEB)* 33(3): 55–63.

Islam, S. N. and A. Gnauck 2011. *Water Shortage in the Gorai River Basin and Damage of Mangrove Wetland Ecosystems in Sundarbans, Bangladesh*. Paper Presented at the 3rd International Conference on Water & Food Management (ICWFM-2011), Dhaka, January 8–10.

Islam, T. and P. Atkins 2007. Indigenous Floating Cultivation: A Sustainable Agricultural Practice in the Wetlands of Bangladesh. *Development in Practice* 17(1): 130–136.

Johnston, B. R. 2010. An Anthropological Ecology? Struggles to Secure Environmental Quality and Social Justice. *Kroeber Anthropological Society* 101(1): 3–21.

Kapoor, D. 2011. Adult Learning in Political (un-civil) Society: Anti-Colonial Subaltern Social Movement (SSM) Pedagogies of Place. *Studies in the Education of Adults* 43(2): 128–146.

Karim, A. 2004. Implications on Ecosystems in Bangladesh. In *The Ganges Water Dispersion: Environmental Effects and Implications.* M. M. Q. Mirza ed. Pp. 125–161. Netherlands: Kluwer Academic Publishers.

Khadka, A. K. 2010. The Emergence of Water as a 'Human Rights' on the World Stage: Challenges and Opportunities. *Water Resources Development* 26(1): 37–49.

Khan, T. R. 1996. Managing and Sharing of the Ganges. *Natural Resources Journal* 36: 456–479.

Khush, G. S. 2001. Green Revolution: The Way Forward. *Nature Reviews Genetics* 2: 815–822.

Kirby, P. 2006. Theorising Globalization's Social Impact: Proposing the Concept of Vulnerability. *Review of International Political Economy* 13(4): 632–655.

Klawitter, S. and H. Qazzaz 2005. Water as Human Right: The Understanding of Water in the Arab Countries of the Middle East. *Water Resources Development* 21(2): 253–271.

Kottak, C. P. 1999. The New Ecological Anthropology. *American Anthropologist* 101(1): 23–35.

Langford, M. 2005. The United Nations Concept of Water as a Human Right: A New Paradigm for Old Problem? *Water Resource Development* 21(2): 273–282.

Latour, B. 1987. *Science in Action: How to Follow Scientists and Engineers through Society.* Cambridge, MA: Harvard University Press.

———— 1995. *Conversations on Science, Culture, and Time.* R. Lapidus, trans. Michigan: The University of Michigan Press.

———— 2004. *Politics of Nature: How to Bring the Sciences into Democracy.* C. Porter, trans. Cambridge and London: Harvard University Press.

Leach, M., R. Mearns and I. Scoones 1999. Environmental Entitlements: Dynamics and Institutes in Community-Based Natural Resource Management. *World Development* 27(2): 225–247.

Lebel, L., P. Garden and M. Imamura 2005. The Politics of Scale, Position, and Place in the Governance of Water Resources in the Mekong. *Ecology and Society* 10(2): 1–19.

Lewellen, Ted C. 1992. *Political Anthropology: An Introduction,* Westport, CT: Bergin and Garvey.

Ludden, D. 2001. Subalterns and Others in the Agrarian History of South Asia. In *Agrarian Studies: Synthetic Work at the Cutting Edge.* J. C. Scott and N. Bhat eds. Pp. 205–235. Yale: The Yale ISPS Series.

Makki, F. 2004. The Empire of Capital and the Remaking of Centre-Periphery Relations. *Third World Quarterly* 25(1): 149–168.

Mann, G. 2009. Should Political Ecology be Marxist? A Case for Gramsci's Historical Materialism. *Geoforum* 40(3): 335–344.

Martinez-Alier, J., G. Kallis, S. Veuthey, M. Walter and L. Temper 2010. Social Metabolism, Ecological Distribution Conflicts, and Valuation Languages. *Ecological Economics* 70: 153–158.

Mauss, M. 1990. *The Gift: The Form and Reason for Exchange in Archaic Societies.* London: Routledge.

McGregor, J. 2000. *The Internationalization of Disputes over Waters: The Case of Bangladesh and India.* Paper Presented at the Australasian Political Studies Association Conference, Canberra, October 3–6. http://apsa2000.anu.edu.au/confpapers/mcgregor.rtf, accessed May 3, 2012.

Mehta, L. 2005. Unpacking Rights and Wrongs: Do Human Rights Make a Difference? The Case of Water Rights in India and South Africa. *Institute of Development Studies Paper 260.* Sussex: Institute of Development Studies.

M'Gonigle, R. M. 1999. Ecological Economies and Political Ecological: Towards a Necessary Synthesis. *Ecological Economics* 28: 11–26.

Mirumachi, N. and J. A. Allan 2007. Revisiting Transboundary water Governance: Power, Conflict, Cooperation and the Political Economy. *Proceedings from CAIWA International Conference on Adaptive and Integrated Water Management: Coping with Scarcity.* Basel, Switzerland, November 12–15.

Mirza, M.M.Q. 1997. Hydrological Changes in the Ganges System in Bangladesh in the Post-Farakka Period. *Hydrological Sciences Journal-des Sciences Hydrologiques* 42(5): 613–631.

Mirza, M.M.Q. and N. J. Ericksen 1996. Impact of Water Control Projects on Fisheries Resources in Bangladesh. *Environmental Management* 20(4): 523–539.

Mirza, M. Q. and M. H. Sarker 2004. Effects on Water Salinity in Bangladesh. In *The Ganges Water Dispersion: Environmental Effects and Implications.* M. Monirul Qader Mirza, ed. Pp. 81–102. Netherlands: Kluwer Academic Publishers.

Moore, J. W. 2000. Sugar and the Expansion of the Early Modern World-Economy: Commodity Frontiers, Ecological Transformation, and Industrialization. *Rev. Fernand Braudel Cent.* 23: 409–433.

Mosse, D. 1997. The Symbolic Making of a Common Property Resource: History, Ecology and Locality in a Tank-Irrigated Landscape in South India. *Development and Change* 28: 467–504.

——— 2013. The Anthropology of International Development. *Annual Review of Anthropology* 42: 227–246.

Nilsson, C., C. A. Reidy, M. Dynesius and C. Revenga 2005. Fragmentation and Flow Regulation of the World's Large River Systems. *Science* 308: 405–408.

Nygren, A. 1999. Local Knowledge in the Environment-Development Discourse: From Dichotomies to Situated Knowledge. *Critique of Anthropology* 19(3): 267–288.

Okongwu, A. F. and J. P. Mencher 2000. The Anthropology of Public Policy: Shifting Terrains. *Annual Review of Anthropology* 29: 107–124.

O'Manique, J. 1992. Human Rights and Development. *Human Rights Quarterly* 14: 78–103.

Orlove, B. and S. C. Caton 2010. Water Sustainability: Anthropological Approaches and Prospects. *The Annual Review of Anthropology* 39: 401–415.

Ostrom, E., J. Burger, C. B. Field, R. B. Norgaard and D. Policansky 1999. Revisiting the Commons: Local Lessons, Global Challenges. *Science* 84, 284(5412): 278–282.

Pamukcu, K. 2005. The Right to Water: An Assessment. *Contemporary Politics* 11(2): 157–167.

Parua, P. K. 2010. *The Ganga: Water Use in the Indian Subcontinent.* Dordrecht: Springer.

Pasi, N. and R. Smardon 2012. Inter-Linking of Rivers: A Solution for Water Crisis in India or Decision in Doubt? *The Journal of Science Policy and Governance* 2(1): 1–42.

Paul, B. K. 1984. Perception of Farmers and Agricultural Adjustment to Floods in Jamuna Floodplain, Bangladesh. *Human Ecology* 12(1): 3–19.

———— 1995. Farmer's Response to the Flood Action Plan (PAP) of Bangladesh: An Empirical Study. *World Development* 23(2): 299–309.

———— 1999. Women's Awareness and Attitudes Towards the Flood Action Plan (PAP) of Bangladesh: An Comparative Study. *Environmental Management* 23(1): 103–114.

Peet, R. and M. Watts 1996b. Liberation Ecologies: Development, Sustainability, and Environment in an Age of Market Triumphalism. In *Liberation Ecologies: Environment, Development, Social Movements*. R. Peet and M. Watts eds. Pp. 1–23. New York: Routledge.

Pels, P. 1997. The Anthropology of Colonialism: Culture, History, and the Emergence of Western Governmentality. *Annual Review of Anthropology* 26: 163–183.

Potkin, A. 2004. Watering the Bangladeshi Sundarbans. In *The Ganges Water Dispersion: Environmental Effects and Implications*. M. Monirul Qader Mirza ed. Pp. 163–176. Netherlands: Kluwer Academic Publishers.

Prebisch, R. 1963. *Toward a Dynamic Development Policy for Latin America*. New York: UN Economic Commission for Latin America.

Rasul, G. 2014a. Food, Water, and Energy Security in South Asia: A Nexus Perspective form the Hindu Kush Himalayan Region. *Environmental Science and Policy* 39: 35–48.

———— 2014b. Why Eastern Himalayan Countries Should Cooperate in Transboundary Water Resource Management. *Water Policy* 16: 19–38.

Robbins, P. 1998. Authority and Environment: Institutional Landscapes in Rajasthan, India. *Annals of the Association of American Geographers* 88(3): 410–435.

———— 2012. *Political Ecology: A Critical Introductions*. Malden, MA: Wiley-Blackwell.

Sachs, W. 1992. *The Development Dictionary: A Guide to Knowledge as Power*. London: Zed Books.

Scott, J. C. 1985. *Weapons of the Weak: Everyday Forms of Peasant Resistance*. New Haven and London: Yale University Press.

———— 1998. *Seeing like a State: How Certain Schemes to Improve the Human Condition Have Failed*. New Haven and London: Yale University Press.

Selby, J. 2007. *Beyond Hydro-Hegemony: Gramsci, the National, and the Trans-National*. Paper Presented at the 3rd International Workshop on Hydro-Hegemony, London School of Economics. http://www.soas.ac.uk/water/publications/papers/file39697.pdf, accessed May 12, 2013.

Shiva, V. 2005. *Earth Democracy: Justice, Sustainability and Peace*. Cambridge, MA: South End Press.

Sillitoe, P. 1998. What, Know Natives? Local Knowledge in Development. *Social Anthropology* 6(2): 203–220.

———— 2000. Let Them Eat Cake: Indigenous Knowledge, Science and the 'Poorest of the Poor'. *Anthropology Today* 16(6): 3–7.

Skogly, S. I. 1993. Structural Adjustment and Development: Human Rights – An Agenda for Change. *Human Rights Quarterly* 15: 751–778.

Sobhan, R. 1981. Bangladesh and the World Economic System: The Crisis of External Dependence. *Development and Change* 12(3): 327–347.

Stone, I. 1985. *Canal Irrigation in British India: Perspectives on Technological Change in a Peasant Economy*. Cambridge: Cambridge University Press.

Sultana, F. and A. Loftus eds. 2012. *The Right to Water: Politics, Governance and Social Struggles*. London: Earthscan.

Swain, A. 1996. Displacing the Conflict: Environmental Destruction in Bangladesh and Ethnic Conflict in India. *Journal of Peace Research* 33(2): 189–204.

Torry, W. I., A. A. William, D. Bain, H. J. Otway, R. Baker, F. D'Souza, P. O'Keefe, J. P. Osterling, B. A. Turner, D. Turton and M. Watts 1979. Anthropological Studies in Hazardous Environments: Past Trends and New Horizons (and comments and reply). *Current Anthropology* 20(3): 517–540.

Toye, J. 1987. *Dilemmas of Development: Reflections on the Counterrevolution in Development Theory and Policy*. Oxford: Basil Blackwell.

Turner, T. 1997. Human Rights, Human Difference: Anthropology's Contribution to an Emancipatory Cultural Politics. *Journal of Anthropological Research* 53(3): 273–291.

United Nations (UN) 1948. Universal Declaration of Human Rights. Available at http://www.un.org/en/universal-declaration-human-rights/. Accessed January 1, 2014.

—— 2002. General Comment No. 15 (2002): The Right to Water. http://www2.ohchr.org/english/issues/water/docs/CESCR_GC_15.pdf, accessed June 23, 2010.

—— 2002. The Right to Water and Sanitation: Articles 11 and 12 of the International Covenant on Economic, Social and Cultural Rights. Geneva: UN.

Van der Pijl, K. 1998. *Transnational Classes and International Relations*. London: Routledge.

Van Ufford, P. Q. 1993. Knowledge and Ignorance in the Practices of Development Policy. In *An Anthropological Critique of Development: The Growth of Ignorance*. Mark Hobart ed. Pp. 135–160. London: Rutledge.

Vayda, A. P. and B. J. McCay 1975. New Directions in Ecology and Ecological Anthropology. *Annual Review of Anthropology* 4: 293–306.

Verghese, B. G., R. R. Iyer, Q. K. Ahmad, B. B. Pradhan and S. K. Malla 1994. *Converting Water into Wealth: Regional Cooperation in Harnessing the Eastern Himalayas Rivers*. New Delhi: Konark.

Wagner, J. 2009 Water Governance Today. *Anthropology News* 51(1): 5, 9.

—— ed. 2013. *The Social Life of Water*. New York: Berghahn.

Wallerstein, I. 2004. *World System Analysis: An Introduction*. Durham and London: Duke University Press.

Wikimedia 2013. Ganges. http://upload.wikimedia.org/wikipedia/commons/3/34/Ganges-Brahmaputra-Meghna_basins.jpg, accessed December 13, 2013.

Wolf, A. T. 1997. International Water Conflict Resolution: Lesson from Comparative Analysis. *Water Resources Development* 13(3): 333–365.

Wood, G. 1999. Contesting Water in Bangladesh: Knowledge, Rights and Governance. *Journal of International Development* 11: 731–754.

Wood, G. D. 1984. Provision of Irrigation Services by the Landless – An Approach to Agrarian Reform in Bangladesh. *Agricultural Administration* 17: 55–80.

World Commission on Environment and Development 1987. Our Common Future. http://conspect.nl/pdf/Our_Common_Future-Brundtland_Report_1987.pdf, accessed May 12, 2014.

Zaman, M. Q. 1993. Rivers of Life – Living with Floods in Bangladesh. *Asian Survey* 33(10): 985–996.

2 The Ganges dependent area in Bangladesh

In this chapter, I describe the basin hydrological and ecological characteristics of the Ganges Dependent Area (GDA) in which Chapra is located, noting how government interventions such as the Farakka Barrage and the Flood Action Plan have undermined the ability of the GDA to provide essential ecosystem services to the majority of its residents. Traditional agricultural livelihoods in this region are well adapted to seasonal variations in rainfall and to regular episodes of *borsha* (wet) and *khora* (dry), which produce a variety of freely available ecological services. These services include river fisheries, the seasonal availability of wild foods, animal forage, soil replenishment through siltation and construction materials, such as wood and bamboo. The political culture and centralized political structure of Bangladesh allows elites at the local, national and international levels to establish development programs that give them control over natural resources that promote scientific and technological knowledge at the expense of local knowledge, and promote the privatization and commodi-fication of water and agricultural resources. Therefore, instead of a predictable and manageable variation of rainfall during each season, farmers now face more extreme and less predictable episodes of *bonna* (extreme flooding) and drought. The data I present in this chapter demonstrates that the current government approach is not sustainable when evaluated in relationship to the ecosystems and services on which the majority of households depend for their livelihoods.

The dependency of the Gorai bank communities on Ganges hydro-ecology

Bangladesh is a riverine country with more than 94 percent of its river water originating from international rivers of India. The Ganges Dependent Area (GDA) occupies the entire southwest area of Bangladesh and is classified by government agencies as one of eight major hydrological zones in the country (Figure 2.1). The southwest hydrological zone is dependent on the water that originates from as far as 2,600 kilometers away in the Himalayan Mountains and flows through the four countries of China, Nepal, India and Bangladesh before reaching the Bay of Bengal. The Ganges Basin as a whole has a

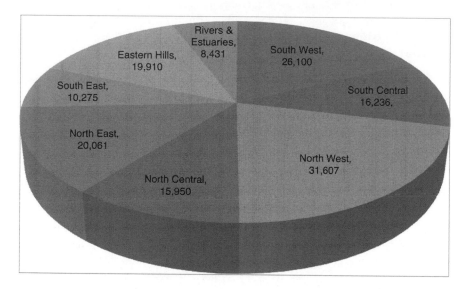

Figure 2.1 The major hydrological zones in Bangladesh (in square kilometers)
Source: Ministry of Water Resources (2001:28)

catchment area of 33,520, 147,480, 860,000 and 46,300 square kilometers respectively within each of these countries (Ministry of Water Resources n.d.). In this study, I focus on the GDA but more specifically on the Gorai River, one outlet of the Ganges in the southwest zone, and on Chapra village in Kushtia, which is located alongside the Gorai.

The GDA in southwestern Bangladesh includes river floodplains, peat basins, estuaries and the mangrove-dominated Sundarbans. The mangrove forests of the Sundarbans depend on the Ganges Basin freshwater flow from the Gorai River and provide support for a unique regional ecosystem. The Sundarbans are one of the largest mangrove regions of the world and are dominated by freshwater swamp forests.

About 80 percent of Bangladeshi people, including those who live in the GDA, live in rural areas and are dependent on water and agricultural practices for their livelihoods. In communities like Chapra, the rainy season brings regular water flow, locally called borsha, from the Ganges Basin. The borsha or wet season transports siltation onto agricultural lands that is helpful for promoting soil fertility and agricultural production. Local farmers do not need to use chemical fertilizer because of this natural fertility source. This is the most productive time of the year for both crops and for other natural resources, like the water-borne wild vegetables, fisheries and vegetation that are also a major foundation of livelihood practices. Local people do not need to buy these from markets as they are freely available from wild sources. Before the borsha season, local communities get khora, or a dry season, that is helpful for producing different crops and receiving different wild-vegetables, fish and fruits. However, the Farakka Barrage in India is controlled in such a way that extra water is released during the rainy season which causes *bonna*, or flooding, and drives away a regular borsha season. Khora is also made much worse during the dry, pre-monsoon spring season when water is diverted for use in India and the downstream flow to Bangladesh is drastically reduced (see chapter three for a more detailed account of this issue). The national government of Bangladesh has historically failed to secure a proper share of the basin flow due to power differences with India; to compensate, they have developed a technology-based water and agriculture management system which I describe in chapters three and four. However, this technology-driven system has created further socioecological vulnerabilities for the marginalized basin communities at the southwest hydrological region in Bangladesh, including Chapra.

The Gorai River is the major distributary of the Ganges River in Bangladesh. This river originates in Kushtia District and provides ecosystem services like fresh water to the southwest hydrological region and has protected the Sundarbans mangrove forests from salinity intrusion for about 500 years (Islam and Gnauck 2011). The Gorai bank communities at Chapra are 5 kilometers away from the Ganges River and are entirely dependent on the Ganges and Gorai Rivers for protecting agricultural practices and livelihoods.

The Gorai River flow area is 199 kilometers in length with a catchment area of 15,160 square kilometers that include the districts of Pabna, Chuadanga, Kushtia, Rajbari, Faridpur, Gopalganji, Jessore, Jhenaidah, Magura, Narail, Perojpur, Barguna, Bagerhat, Khulna and Satkhira. Islam and Gnauck (2011) describe five different sections of the Gorai River: the Gorai-1 to Gorai-4 sections include the non-tidal watercourses and the Gorai-5 is the tidal lower watercourse. Both of the tidal and non-tidal sections of the Gorai River discharge into the Bay of Bengal via the Madhumati and Baleswar Rivers.

The Ganges Basin provides major water services for Chapra households including my case study respondents Tanvir, Tofajjal, Joardar and Billal who represent the rich, intermediate, small and landless laborer respectively. According to Joardar, the basin ecosystem provides "*pukur vora mach, gola bhora dhan, goal bhora goru*" (a pond full of fish, a household store room full of rice paddy and a domestic animal shelter full of animals). This basin flow provides ecosystem support to local water bodies like the *Chapaigachi haor* (oxbow lake), Chapra and Lahineepara *beels* (wetlands), Lahineepara and Shaota *khals* (canals) at Chapra. These waterbodies are naturally developed based on the Ganges-Gorai River flow. The borsha flow transports water from the basin to these water bodies and ensures a healthy environment for the whole year. Based on this flow, the better-off farmers, Tofajjal and Tanvir, produce commercial agricultural crops; and the marginalized farmers, Billal and Joardar, grow some of their own food and receive employment opportunities and ecosystem services. Parvin, wife of Billal, and Suma, wife of Joardar, use local water bodies for domestic activities. As noted previously, the basin flow is the prerequisite for basic livelihood activities like producing crops and catching fish but is also essential to a range of social activities such as sailing boats, visiting relatives, organizing water sports and providing resources for seasonal celebrations. Irrespective of their socioeconomic conditions, everybody at Chapra has equal rights to use the basin flow and ecological resources without any discrimination or conflict. In Bangladesh, local communities enjoy these historical rights to river water and resources without any fee or restriction. Harun and Kusum stated: *Nodir paanir madya anek sompod lukiyea thake ja maati ebong amader shairer jonno dorkar. Paani nei mane amader jibon songkote.* (River water has hidden assets essential for earth and our body. Water lost means existence crises.)

They understand the seasonal dynamics of winter, summer and rainy seasons and their connection with the Ganges Basin flow. Based on these seasonal patterns, they practice the three major cropping patterns of *kharif*-1, *kharif*-2 and *robi* for agricultural production (see Table 2.1). Kharif-1 occurs during the period from March to May, from mid-spring to mid-summer. Kharif-2 occurs during the

Table 2.1 Bengali calendar, seasons and cropping periods

Months – Western Calendar	Western Seasons	Months – Bengali Calendar[1] (dates are for the current year)	Bengali Seasons	Cropping periods
January	Winter	*Poush* (Dec.17–Jan.15)	*Sheet*	Robi
February		*Magh* (Jan.16–Feb.13)		Kharif-1
March	Spring	*Phalgun* (Feb.14–Mar.15)	*Basanta*	
April		*Chaitra* (Mar.16–Apr.14)		
May		*Baishakh* (Apr. 15–May 15)	*Khora or Grisma*	Kharif-2
June	Summer	*Jyaistha* (May 16–Jun. 15)		
July		*Asadh* (Jun.16–Jul.17)	*Borsha*	
August		*Shraban* (Jul.18–Aug.17)		
September	Fall	*Bhadra* (Aug.18–Sept.17)	*Sharat*	
October		*Ashwin* (Sept.18–Oct.18)		Robi
November		*Kartik* (Oct.19–Nov.17)	*Hemanta*	
December	Winter	*Agrahayan* (Nov.18–Dec.16)		

1 The Bengali calendar was created during the reign of the Mughal Emperor, Akbar, in the sixteenth century. It was introduced as a replacement for the Islamic Hijra calendar which did not match the harvest seasons in Bengal and was therefore less useful for taxation purposes. The Bengali calendar combines lunar and solar elements and is considered to begin in mid-April with the *Baishakh* season. All Muslims and Hindu religious and cultural practices and holidays are incorporated into this calendar year. The year 2014 in the western calendar is counted as year 1421 in the Bengali calendar.

period from June to October, from the rainy or monsoon season to autumn. The robi cropping season is from November to February, that is, from late autumn, through winter to early spring. The Chapra community cultivates, for example, onion crops during the winter season as a robi crop because this plant cannot be grown during the rainy season. Again, they cultivate aman paddy during the borsha season because this crop requires a higher level of water availability. They developed this system of crop scheduling over several generations and the knowledge it requires creates a deep bond among people, river, water and land.

Marginalized farmers have a deep understanding of the local ecosystem services and use this knowledge to ready crop lands for planting. They visit crop lands immediately after decreases in the borsha flow and evaluate the siltation level. They pick up a piece of soil and test it to understand moisture levels and readiness for starting crop production. They can ready cropland earlier or later by putting water hyacinth on the land. The water hyacinth is a local plant borne in water bodies and can be used to preserve cropland humidity for a longer time. They can also determine what type of crop production is suitable based on this soil type. They cultivate crops like pulse in muddy areas and start other crops based on land conditions.

Billal, Rahim, Laily and Suman explained to me how local land use patterns vary on the basis of cropland elevation. Mid-level lands are naturally developed croplands that differ from the higher and lower croplands. This land can produce single, double or triple harvests in a year depending on the basin flow. Based on the borsha seasonal flow, the higher land can also produce triple crops during kharif-1, kharif-2 and robi. The term *vita* is used to refer to the practice of mounding up soils to overcome challenges of water stagnation and flooding. The term *maacha* is used to refer to another technique for agricultural production that involves construction of an infrastructure on higher elevation lands with bamboo sticks, rope and other local materials; vegetables like cucumbers or potatoes are then planted within this structure. My survey data indicated that 80 percent of the people at Chapra use these techniques to maximize their agricultural production.

The Ganges River is a major source of fertilizer for farmers at Chapra. According to Tofajjal, the regular borsha flow produces siltation and algae in croplands and develops sustainable ecosystem. The borsha flow washes away exhausted top soil from crop fields. Again, the flow produces water hyacinth in local water bodies. Joardar deposits water hyacinths on croplands during the borsha season, as compost, to promote cropland fertility. Also, for almost eleven months in a year, Joardar and his wife, Suma, collect domestic animal dung and decompose it in a pit at his homestead. Joardar transfers this fertilizer onto his croplands immediately after the borsha season for the winter and summer seasons' crop production. Tanvir, the land owner, hires Billal, a traditional day laborer to prepare this fertilizer and hires additional laborers to spread them on his crop fields. Moreover, they receive earth-worm generated fertilizers. This is a long worm locally called *kecho* that burrows into croplands at the beginning of rainy season; the burrowing mechanism improves cropland fertility.

Joardar, Suma and Tanvir provided me with elaborate descriptions of how they produce their own seeds based on local ecosystems. Joardar scrutinizes a specific cropland's elevation, temperature and soil quality for producing crops like rice, onion, ginger and turmeric. He does not need to spend much money to plant these crops since he is able to produce the seeds himself through careful application of his own local knowledge. The female household members like Suma cultivate local crops like bean or watermelon. She also contributes by preserving some of the seeds from these crops. Joardar and Suma demonstrate deep ecological knowledge about temperature and humidity through their seed preservation practices. They preserve onion seeds in an earthenware jar, or *kolosh,* to ensure an ecologically friendly environment that allows no extreme hot or cold. Suma inspects temperature and seed condition every 15 days. As a rich farmer, Tanvir supervises seed production and, Sohali, wife of Tanvir, supervises seed preservation. Based on his supervision, day laborers harvest the seeds when they are ready. After bringing the seeds at home, Sohali supervises her maid servants who preserve them in a suitable place following local knowledge. During cropping schedule, sometimes, they exchange these seeds with trusted neighbors or relatives and develop community bonds.

Chapra community people husband domestic animals like cattle, hens, chickens, ducks, goats and sheep based on river-dependent ecosystem services. Suma and her daughters, Amina, Joshna and Shifaly, look after domestic animals like chickens and ducks. Suma puts them in crates every evening and brings them out every morning. They send ducks to the Chapra *beels*, or wetlands, every day. Joardar's sons Jakir and Karim take care of cows and calves, feeding them local plants. Tanvir hires Billal to take care of his domestic animals. Moreover, Tanvir's domestic help, Parvin, takes care of these animals under the supervision of Tanvir's wife, Sohali. During the borsha season, local people at Chapra use stored paddy straw, water hyacinths, banana plants and bamboo leaves as fodder for domestic animals like bullocks or goats. Based on these efforts, they regularly get eggs and meat from chickens and ducks and produce young chicks every four months. They cultivate croplands with bullocks and get fertilizer, biogas and fuels from these domestic animals. The cows provide milk every day and produce calves every year. They sell extra domestic animals, eggs or milk, which helps to pay for school fees and household items. Joardar gets bullocks from their domestic cows and successfully produces agricultural crops. They use cattle bones, horns and hooves for making agricultural materials like plows, for building houses and for healing purposes.

Billal, Parvin and Kabir informed me that the occupational groups like boatmen, fishermen, blacksmiths, potters, thatchers and basket makers also depend on local agricultural production. The boatmen transport agricultural goods and services, fishermen provide fish and blacksmiths make agricultural materials like hoes for farmers. The potters make household cooking materials and children's toys. The thatcher builds homes for community members. The basket makers provide different types of baskets for community people. Parvin uses these baskets for gardening and collecting fuel woods. Much of this local knowledge transmits from one generation to the next. Children from their own families begin learning this knowledge in early childhood when they see older generations' occupational practices. Grandfathers tell many stories to grandchildren for teaching knowledge about seasonal patterns, local cropping practices, agricultural production materials, housing and transportation practices. According to Billal, young people would traditionally become knowledge experts themselves by the age of 20.

My focus group respondents Rahim, Parvin, Suma and Soheli specified that local natural resources are required to produce a variety of essential agricultural materials. They use bullock carts locally called *gorur garee* for transporting field crops during the summer season. This bullock cart is made with local natural resources like bamboo, wood and jute. During the rainy season, Joardar is used to make a boat with locally available natural resources like trees, bamboo and *gab,* coating materials for transportation. When field crops are harvested, Suma contributes by paddy husking and winnowing. As a marginalized woman, Parvin works at Tanvir's house under Sohali's supervision to winnow the rice plant, boil them and make rice. The winnowing fan is locally called *kula* and is made with bamboo. The paddy husking machine is locally called *deki,* or "husking pedal." They make ploughshares, frames, ladders and sticks from freely available ecological resources like bamboo and trees. They also make other agricultural materials like hoes, sickles and cleavers from the same ecological resources.

Wild vegetables are a traditional ecological resource in Chapra. Some of the wild vegetables are water lilies, marsh herbs, water spinaches, hyacinth beans and ferns available at the local water bodies like Chapra and Lahineepara beels. Parvin, wife of Billal, regularly collects water lilies and their fruits for family consumption. Sometimes, their children Jamal, Lablu, Kamrun and Shampa collect them from these water bodies. During the times of food shortages, they collect and eat water lily fruits to overcome starvation. Wild banana trees are also readily available close to their homesteads, in forested areas and along the Gorai River banks. They also eat wild bananas from roadside trees and nobody complains about these practices. They collect midribs and inflorescences from the wild banana trees for household vegetable consumption and also the leaves and roots of arum plants. In addition to these vegetables, Billal gets other vegetables like jute leaves, onion flowers and pulse leaves from Tanvir's crop fields for household consumption.

The borsha season also provides more than sufficient fisheries in local water bodies like Chapra and Lahineepara beels. These fisheries freely accessible to all family and community members promote bonds by sharing them. For example, Billal's father caught a nine-kilogram carp fish and distributed the major portion to neighbors. Billal informed me that this fish sharing develops community bonds and overcomes the risk of loss through rotting. This is a common practice for Chapra communities. Billal is welcome to catch fish in Tofajjal's ponds like the other neighbors. Tanvir prepares ponds using a traditional mechanism to shelter the different types of fish. He puts water hyacinths at a side of the pond so that, for example, catfish and snakehead fish can find suitable habitat. Another side of the pond is covered with water lilies for the fish like carp. The community people preserve local species alive, like catfish, after catching them from local water bodies. They keep these fish in earthenware jars. Suma takes care of these reserved fish and cooks some every week for household consumption. The Ganges Basin ecosystems are a major source of these fish, which are a core feature of socio-economic foundations at Chapra. However, these ecosystems are being profoundly disrupted by the Farakka Barrage diversion in India. Harun and Marjina, two focus group respondents, stated: *"nodirpaani shunnota lobonaktota barai, shoshway utpadan, maas, gokhaddoy ebong matir utpadhan khomota komay"* (River flow failure causes salinity problem and major reduction of crop production, wild fisheries, fodder and land fertility).

The Farakka Barrage

The Ganges Basin countries have failed to develop effective basin management agreements and have thereby created major ecosystem failures and livelihood challenges for basin communities such as Chapra. The central government in India was able to build the Farakka Barrage unilaterally in 1975 because of their hydropolitical dominance over the region (Turton and Henwood 2002; Zeitoun and Warner 2006). The barrage was constructed mainly in order to divert additional water into the Hoogly River, a branch of the Ganges that flows through Calcutta and empties in the Bay of Bengal. The additional water ensures that the port of Calcutta can remain open year round rather than suffer closures due to sedimentation. India has also built a small hydroelectric facility at the barrage and makes some of the stored water available for industrial and irrigation uses within West Bengal (Iyer 1997:4; Khan 1996; Swain 1996). Iyer (1997:4) notes that

> the primary purpose of the Farakka Barrage was the diversion of a part of the waters of Ganges to the Bhagirathi/Hooghly arm to arrest the deterioration of Calcutta Port. The secondary purpose was to protect Calcutta's drinking and industrial water supplies from the incursion of salinity.

The central government in Bangladesh was not able to prevent the construction of this barrage due to disparities of geographic size, economic capability and political and military strength. The governments of India and Bangladesh have concluded two water treaties, one in 1977 and another in 1996 and two Memoranda of Understanding, in 1983 and 1985. The goal of the most recent treaty, the Ganges Treaty of 1996, signed for a thirty-year period, is to enable water sharing of the Ganges River, especially during the dry season. However, the treaty does not address the specific ecological concerns reported above regarding borsha failures, bonna, drought, river bank erosion, water stagnation and embankment failure. And, despite its stated intent, the 1996 Ganges Treaty has also not resulted in positive outcomes for khora (dry season) flow, as I explain below.

Ganges River flow data measured in Bangladesh at the Hardinge Bridge Station, six kilometers distance from the mouth of the Gorai River and eleven kilometers distance from Chapra, indicates how the basin flow changes during the rainy and dry summer seasons (Table 2.2 and Figure 2.2). During the

Table 2.2 Ganges River flow data before and after construction of the Farakka Barrage

Flow rates at Hardinge Bridge in Bangladesh in cubic feet per second (cusecs)

Year	Khora (March–April) seasonal flow (cusecs)					Borsha (July–September) seasonal flow (cusecs)				
	Average	Max	Month	Min	Month	Average	Max	Month	Min	Month
1960	2326	2480	Mar 07	2170	Apr 30	39794	48000	Sept 04	25800	Aug 02
1964	2320	2600	Mar 01	2180	Apr 15	39733	48300	Sep 10	29200	Jul 28
1970	2360	2640	Mar 11	2030	Apr-24	32917	40800	Aug 18	24400	Aug 30
1976	780	1130	Mar 01	657	Mar 29	33569	50000	Aug 31	19100	Jul 31
1980	927	962	Mar 02	874	Mar 30	48350	57800	Aug 22	37500	Aug 02
1985	823	1020	Mar 05	701	Apr 04	37373	48000	Aug 29	24200	Jul 25
1990	828	1030	Apr 29	698	Mar 01	40444	51000	Aug 21	23100	Sep 09
1995	509	769	Mar 01	363	Apr 26	35219	48800	Aug 19	18800	Jul 30
2001	766	997	Apr 27	456	Apr 21	34390	44004	Aug 30	22095	Aug 19
2006	828	1092	Apr 07	418	Apr 22	26271	35079	Aug 31	21887	Aug 15
2010	743	970	Mar 20	475	Mar 31	25190	40276	Sep 25	8701	Jul 05

Source: Bangladesh Water Development Board (2012)

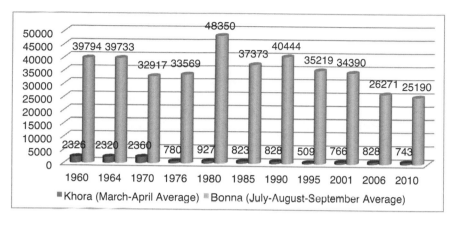

Figure 2.2 Average Ganges flow before and after the Farakka Barrage

Source: Bangladesh Water Development Board (2012)

summer season the flow of water in the Ganges is reduced significantly due to less rainfall and flow reductions at its place of origin in the Himalaya Mountains. The Farakka Barrage authority diverts enough flow for India's needs into the Hoogly River during the dry season but is able to limit the flow of water into the Hoogly during the wet season. This limits flooding in West Bengal but worsens it in Bangladesh. Table 2.2 shows that the average daily water flow into Bangladesh during the dry season has been reduced two-thirds since the Farakka Barrage began operating in 1975 (see also Figure 2.2). Currently, the Gorai River is at the verge of extinction due to this basin flow reduction.

Joardar and Billal argued that reductions to river flows during the borsha season are the most common cause of crop production and employment failures for Chapra. The borsha season occurs every year during the months of July, August and September, with water levels reaching their peak in August. For this study, I calculated the average flow from July 25 to September 10 every year for the last 50 years, from 1960 to 2011, to demonstrate historical changes of flow during this season. Table 2.2 data demonstrates that the minimum flows are significantly lower on average during the rainy season. This reduction causes failures of aman rice production, a staple crop on which households depend for the whole year and which cannot mature without sufficient water. This seasonal flow reduction also causes ecological service failures of the kinds described above.

The khora season occurs during the months of February, March, April and May every year. It has its own distinct growing season, which begins in January and ends in May; water scarcity begins in February, reaches a peak in April and ends in May. Based on the months of March and April, I calculated an average flow for the river every year for the khora season. This flow provides the basis for understanding water crises and their consequences for dry season agricultural production, employment and livelihood practices.

Table 2.2 indicates major discrepancies in the Ganges flow before and after construction of the Farakka Barrage. In 1960, the average flow was 2,326 and 39,794 cubic feet per second (cusecs) during the summer and rainy seasons respectively. There was no major deviation of the flow for the next 14 years from 1960 to 1974. In 1974, for example, the average flow was 2,393 and 43,369 cusecs during the summer and rainy seasons respectively. However, the average flow dropped as low as 780 cusecs during the summer season in 1976, immediately after the Farakka Barrage began to divert water in India (Table 2.2). This flow reduction created a sudden water shock to ecosystem in the GDA and subsequently to agricultural production and employment in the GDA in Bangladesh. The Ganges River flow has never returned to pre-Farakka levels and has, in fact, encountered increasing reductions over the last 40 years. In 2010, the average basin flow was 743 and 25,190 cusecs per day during the summer and rainy seasons respectively (Table 2.2).

Based on the 1996 treaty, Bangladesh should receive more than 35,000 cusecs water in every alternate 10-day period during the summer season depending on the basin flow conditions (Table 2.3). However, in the dry season, when flows are generally much lower than 70,000 cusecs, other provisions of the treaty are applied. Article-II (iii) informs us that in case of the river flow reduction below 50,000 cusecs in any 10-day period at the Farakka point, the governments of India and Bangladesh will work together on an emergency basis to ensure "equity, fair play and no harm" to either party (Government of Bangladesh 1996). However, when comparing the pre-Farakka flow rates to the most recent flow rates for the dry season, it is apparent that Bangladesh now receives only about one third of the former flow, suggesting that India diverts two thirds of the dry season flow for their own use. Furthermore, the major limitation to measuring the specific amount of the diversion is that the Joint River Commission fails to publish the Ganges Basin flow data at the Farakka Barrage point.

In addition to problems associated with the enforcement of its water sharing provisions, the treaty fails to describe guidelines for ecological concerns associated with borsha seasonal failures, flooding, river bank erosion and embankment

Table 2.3 Annexure 1 of the 1996 Ganges Treaty for sharing Ganges River flow

Availability at Farakka	Share of India	Share of Bangladesh
70,000 cusecs or less	50%	50%
70,000 cusecs – 75,000 cusecs	Balance of flow	35,000 cusecs
75,000 cusecs or more	40,000 cusecs	Balance of flow

Source: Government of Bangladesh (1996)

failures (Bhattarai 2009; Brichieri-Colombi and Bradnock 2003:53; Gupta et al. 2005; McGregor 2000; Rahaman 2009). Article VIII focuses on the mechanisms of cooperation for the different issues like flood management, irrigation and hydropower based on the principles of "equitable" utilization and "no harm" that are described in articles IX and X (Government of Bangladesh 1996). Article IX extends these principles to other international rivers shared by India and Bangladesh; however, to date, none of the river disputes between the countries have been resolved, including that over the Teesta River, which has been a major source of contention (Padmanabhan 2014; PTI 2014). The government of India has also ignored the principles of the Ganges River Treaty by unilaterally promoting their National River Linking Project (NRLP) that involves massive water transfer from the Brahmaputra to the Ganges Basin to serve industrial growth in drought prone areas in India. The implementation of NRLP is likely to bring about ecological disasters in the rest of Bangladesh similar to those currently being encountered in the southwest (Bandyopadhyay and Ghosh 2009:54; Bhattarai 2009:4; Haftendorn 2000:64; Khalequzzaman 1994; Swain 1996).

Despite the existence of the treaty, the basin flow continues to diminish in Bangladesh. Shortly after the treaty was signed in 1996, the average flow diminished from 828 cusecs in 1990 to 766 cusecs in 2001 during the dry season (Table 2.2). In 1990, prior to signing the treaty, the average flow during the khora season was 828 cusecs, down from a slightly higher level in 1988 of 1,011 cusecs. The average flow fluctuated between 1,058 and 866 cusecs from 1998 to 1999 but by 2001 it had diminished to 766 cusecs. In 2005, the average flow remained low at 782 cusecs and then declined further in 2010 to 743 cusecs. Rainy seasonal flows since the signing of the treaty in 1996 also show a consistent reduction. In 1995, the basin flow was 35,219 cusecs during the rainy season but this diminished to a historic low of 25,190 cusecs in 2010 (Table 2.2). The 1996 Ganges Treaty has clearly failed to achieve positive outcomes for Ganges water sharing between India and Bangladesh.

Flow records for the borsha season since 1960 also indicate an increased incidence of bonna or severe flooding. Records for the 1960 to 1970 period indicate that maximum flows ranged from 40,800 to 48,000 cusecs. However, in 1976, 1980 and 1990, the maximum flows ranged from 50,000 to 57,800 cusecs. The unpredictability of water flows is as damaging for basin communities as either floods or shortages since it makes impossible to effectively plan ahead for agricultural production.

The unpredictability of water flows also has direct ecological impacts. In 1982, the summer season average flow was 1,426 cusecs whereas this flow diminished to a low of 845 cusecs the very next year, in 1983 (Bangladesh Water Development Board 2012). The average flow reached to a low of 558 cusecs in 1989 and increased to 828 in 1990. Again, this average flow shrank to 358 cusecs in 1993 and increased to 1,058 cusecs in 1998 (Bangladesh Water Development Board 2012). These flow fluctuations generate sedimentation throughout the basin and in local water bodies as water flow loses its capacity to carry sedimentation at downstream and distribute it in an unpredictable manner. On the other hand, variations in the rainy seasonal flow, especially those associated with extreme floods, cause severe river bank erosion and embankment failures. Most Chapra households have no means to overcome or mitigate these concerns and face major damage and loss of crop production, household infrastructure and agricultural materials.

While a certain range of fluctuation was normal before the construction of the Farakka diversion, the Farakka operation has magnified these fluctuations to a destructive and unmanageable levels. Sometimes, the length of the khora season is shorter or longer than the regular season due to domination of the Farakka diversion. Sometimes, the khora season begins in December or January and ends in March or July. For example, the khora season began on 31 March in 2009 with the flow of 602 cusecs and ended on 25 May with the flow of 948 cusecs. In 1979, the khora season began in January and continued about six months till June. The basin flow was 1,520 and 1,090 cusecs on January 29 and June 18 respectively in 1979, which caused major changes to the summer season. In 2009, the khora season began in January and ended in June (Bangladesh Water Development Board 2012). Again, this fluctuation occurs during the borsha season as well as the khora season. Currently, there is no regular borsha season and it gets longer or shorter depending on the Farakka diversion from the Ganges Basin flow.

These ecosystem failures create major challenges for community sustainability in the GDA. They also pose significant challenges to the political and institutional culture of Bangladesh as it seeks to mitigate these problems while pursuing its own development agenda. As I explain below, the political culture of Bangladesh and the policy framework within which they attempt to manage water resources, tends to replicate and exacerbate, rather than mitigate, the problems created by India through construction of the Farakka Barrage.

Political culture

The Peoples' Republic of Bangladesh is a parliamentary democracy similar in structure to many other British Commonwealth nations. Unlike Canada and India, however, it is not divided into Provinces or States that each elects their own legislative assemblies. It is divided into eight administrative divisions and those divisions are further sub-divided into 64 districts. Local elections are held to populate some of the positions on sub-district level (upazila) councils and the union councils that represent a group of villages, but other members of these councils are appointed by the central government. There are 64 District Councils, 483 Upazila Parishads and 4,500 Union Councils in Bangladesh (Bangladesh Bureau of Statistics 2010). Below the union level there is also a ward level with nine wards in very Union Council. Every ward in turn is made up of two to four villages. There are no official government institutions at ward and village levels but villages especially organize a large number of events through traditional informal mechanisms. They organize religious events, sporting events and various kinds of meetings to discuss events that might require collective action such as an erosion problem or assistance to a neighbor in need.

Since the authority of upazila, union and village councils is very limited by comparison to that of the central government, and since no formal government exists below the union level, the system is structurally one that allows for a great deal of top-down control. The political culture that has emerged in Bangladesh since independence in 1971 is partly an outcome of the structural characteristics of this system as well as the other historical and cultural factors.

Members of Parliament (MPs) are given appointments to the different ministries like water resources, agriculture and finance depending on their personal 'connection' with the Prime Minister. Senior bureaucrats within each ministry are also appointed by the Prime Minister, from a preference of ideological ground within the governing political party. Ministers will appoint senior bureaucrats and they will also be appointed on the basis of their service within the governing party, and their personal 'connection' to the Minister. The Water Resource Minister thus heads a bureaucratic organization that extends from the central government to the sub-district level and is dominated at all levels by members of the governing party who only retain their positions for as long as they demonstrate loyalty to those above them. Government Ministers thus tend to be insulated within a closed circle of political leaders at national and local levels who mobilize power structures based on their own agendas. These leaders constitute a political-economic elite with very little accountability to the grass-roots interests of local communities like Chapra.

MPs, in addition to having the opportunity to be appointed to a ministerial position, may also be nominated for an advisory position at a local resource management committee at the district and sub-district levels. Based on this official position, they control local power structures based on local natural and government resources. They nominate party candidates for local government elections at Upazila Parishad (UP) or Union Council (UC) levels. These local elections run independently and are held at different times of the year than national elections.

The *Zila Parishads* or District Councils are composed of both elected and appointed members. The elected members include all the MPs from that district, the Upazila Parishad Chairmen, the Union Council Chairmen and the Mayors of all municipalities within the district. The Zila Parshad has equal representation from both urban and rural areas. Urban municipalities are governed by mayors and councils while rural areas are governed by Upazila Parishads and Union Councils. The Chair of the Zila Parishad is appointed by the national government based on political loyalty and the parishad also appoints additional members from the local area, usually members of the local elite. In 2016, the Bangladesh government reviewed the Zila Parishad law and establishes direct election for the chair of the parishad. It is important to review the effectiveness of this new law as political culture remains the same. The national government also appoints a Deputy Commissioner, selected from the Bangladesh civil service, to manage the administrative work of each Zila Parishad. Members of the civil service cannot belong to a political party but are chosen on the basis of their loyalty to the ruling political ideology as well as their formal qualifications.

Upazila Parishads of Sub-district Councils consist of one Chair and two Vice-Chairs who are elected directly to their positions, and all the Union Council Chairs within that sub-district. The Union Councils consist of one Chair and twelve other members who are also elected directly by local citizens. Generally these positions are all filled by elite members of the local communities.

The government party is thus able to control the actions of District Councils by appointing only party supporters to the non-elected positions. This is not the case at the sub-district and union council level but even when a majority of the members of those councils are from opposition parties, they tend not to directly oppose the government over local development activities but rather cooperate as best they can, protecting their class position while waiting for the next national election.

All of the major political parties like the Bangladesh Nationalist Party (BNP) and the Awami League (AL), who currently form the government, have party organizations from the national to the village levels. These parties have top-down hierarchies and the Party Chairperson (or the Prime Minister) dominates the government system. Those closest to the Party Chairperson are appointed to the party's Executive Committee (Ahmed 2013). These executive committee members are political leaders, businessmen, former bureaucrats and military officers. Every Executive Committee member has control over a specific geographic area like a division or districts in Bangladesh. This Executive member dominates their party organizations and election candidate nominations at upazila and union levels. In many cases, this member dominates nominations of Members of Parliament candidate within the geographic area. After a parliament election, the Members of Parliament under the leadership of this Executive member work as a powerful group inside the party and government.

In addition to their extraordinary power over local levels of government, MPs and Ministers are major power brokers for the foreign development agencies and corporations that introduce new technologies into the water and agricultural management system. Billal, as a focus group respondent, described these political leaders as *joke* (blood sucking worms), based on their exploitative activities. These political leaders develop close relationships with military leaders, as well as with civil bureaucrats, professional groups, financiers and fund raisers to promote their own business interests. During elections they behave like business contractors trying to gain control of a particular geographic region or government institution. As one focus group respondent put it: *"voter pore amder keo mone rakhena"* (no elected official like the MP or Upazila Parishad Chairman recognizes our voices after the election). Mallet (2014) in a *Financial Time* article termed these practices of politicians in Bangladesh "as a way of seizing (government) control and making money."

Focus group participants, Sabbir and Tarun, stated that most of these political leaders are urban elites and often expatriate Bangladeshis, living only occasionally in Bangladesh. According to them, these elites are called *bosenter kokil,* or "spring birds," who visit local communities for their own interests. These elites are able to profit by exploiting the Ganges flow reductions, as will be explained in more depth in later chapters. Some, for instance, make money by mining sand from the river at times of low flow and selling the sand to construction industries. The perceptions of my respondents are supported by the fact that more than 50 percent of the MPs in 2009–13 were businessmen (Ahmed 2013). Focus group respondents reported that many MPs do not hesitate to kill local people who raise voices against their interests. They stated, for example, that one MP's son, who is only 18 years old, has killed 12 people to fulfill his family's personal interests. Many of the MPs work as lobbyists and others perform official duties in formulating water and agriculture policies. MPs and senior bureaucrats get opportunities like foreign tours, training and contracts based on this technocentric development approach. Many bureaucrats get assignments as directors to implement large-scale water management projects. Furthermore, government staff get pleasure trip opportunities abroad informally from businessmen and contractors.

Water policies in the GDA

Consistent with the highly centralized nature of Bangladesh government, water policy within the GDA is developed on the basis of a National Water Management Plan (NWMP) and a set of national policy guidelines originally developed in 1999. These guidelines incorporate many of the principles of the Flood Action Plans of the 1990s which were developed in coordination with the external agencies, including the World Bank, United States Agency for International Development (USAID) and United Nations Development Program (UNDP), who provided most of the funds necessary to secure what they promoted as a "permanent" solution to flooding problems in Bangladesh (Custers 1993). This approach is based on the idea that all water management problems can be solved by large-scale engineering projects in combination with neoliberal economic policies that favor the privatization and commodification of water resources and water-dependent economic sectors such as agriculture (Faber and McCarthy 2003; Nuruzzaman 2007). In Bangladesh, as in neighboring India, these projects are often grandiose in scale and work in opposition to local knowledge and traditional systems of agricultural production.

In chapter three, I describe the outcomes of the Gorai River Restoration Project and the Ganges-Kobodak Project, two large engineering projects within the GDA but, in this chapter I wish to focus on the government's approach to policy development, emphasizing the scale and centralized, top-down nature of their planning. Given the negative effects of the Farakka diversion, the national government was especially interested in the GDA; and as a result, the Bangladesh Water Development Board (BWDB), following the NWMP guidelines, executed a feasibility study for an integrated water development project for the region. As a result of this study, BWDB has formulated three major projects (Ministry of Water Resources 2001:306). Firstly, the BWDB has proposed a major dredging project immediately downstream from the Hardinge Bridge on the Ganges River to remove sedimentation and thereby increase the flow of water in northwestern rivers like the Boral and southwestern rivers like Mathabanga that depend on Ganges water. Secondly, they are planning to construct a major barrage on the Ganges River upstream from the headwaters of the Gorai River. This barrage will allow for the storage of water during the borsha season for use later during the winter and spring seasons. Thirdly, in addition to the barrage, they are planning the construction of a headwater structure at the head of the Gorai River. The major purpose of this headwater structure is to control flooding in the GDA during the rainy season and to secure water supplies during the khora season. The headwater structure will be 1,870 meters wide with 84 radial gates, each 18 meters wide that will allow fish passage and controlled boat navigation (Ministry of Water Resources 2001:306). Based on the Ganges Barrage, BWDB has also proposed the construction of two linking channels (channels numbered one and four) that will move water by gravity flow from the Ganges River and the upper reaches of the Gorai River to points further south.

Based on the proposed barrage, link channel one can move water from the Ganges River at a point immediately downstream from the Hardinge Bridge, parallel to the Gorai River west bank, to connect to the Hisna and Mathabanga Rivers. This channel will traverse the existing GK Project infrastructure (described in chapter three) but will be designed so as to avoid damage to it. This channel will move water all the way to the Bay of Bengal by connecting to the Nabaganga-Chitra and Bhairab-Kobadak-Betna Rivers which will reduce salinity problems caused by coastal saltwater intrusion (Ministry of Water Resources 2001:325). The central government also estimates that 7,324 square kilometers of agricultural land can be irrigated from this channel. Link Channel 4 will require construction of a barrage on the Ganges River at Tangorbari or Pangsha, on the east side of the Gorai River. This channel 4 will divert the basin flow to the Chandana River and Madaripur Beel. The government estimates that 3,596 square kilometers of net agricultural area will receive irrigation services from this channel.

In order to implement this ambitious set of goals the government also proposes a significant utilization of the bureaucracy necessary to its planning, construction and operation. BWDB will construct the barrage, headwater structures and link channels, while local governments will build irrigation drainage canals. NGOs will take care of land acquisition and resettlement activities. The Ministry of Local Government and Rural Development established a Central Training Unit for local water managers and the staff of the Local Government Engineering Department and the Department of Public Health and Engineering. The Central Training Unit also trains the central government staff of the Water Resources Planning Organization, the Department of Environment, the Disaster Management Bureau and the BWDB (Water Resources Planning Organization 2001). The plan of action described above also involves an expansion of the responsibilities of local water management institutions at the upazila (sub-district) level. Local institutions currently have authority over areas of 1,000 hectares but their mandates will be expanded to 5,000 hectares. After completion of the projects, the central government will assign local water management responsibilities to community organizations or individual pump owners (Ministry of Water Resources 2001), thereby promoting private sector involvements in water resource management at union and village levels. The problem is that the history of large-scale water management projects in the GDA suggests that the outcomes of this particular project will be very different and much less positive than those envisioned by project planners.

Conclusion

The Farakka Barrage diversion from the Ganges Basin creates major challenges for water and agriculture management in the Ganges Dependent Area (GDA) of Bangladesh. During the summer season, the barrage authority diverts much water from the basin that is responsible for *khora* in this GDA. Local water bodies are dying and creating multiple socioecological effects on local communities. The marginalized people at Chapra fail to ensure fodder for their domestic animals and encounter agricultural production damage and loss. The continuous basin flow reduction describes the increasing ecological vulnerabilities of *borsha, bonna, khora,* river bank erosion, embankment failure and water stagnation for Chapra villagers. The government in Bangladesh fails to deal with India and follows technocentric water and agricultural policies to overcome these effects. These policies fail to include community voices as it follows the top-down approach. This water development approach creates failures of recognizing ecology based livelihood practices of siltation services, water-borne vegetables, fisheries and traditional occupations like fishing and boat driving. Furthermore, this approach adds new concerns of river bank erosion, embankment failure, flood and drought which cause multiple socioeconomic challenges. Based on this argument, I describe the outcomes of two previous, on-going projects, the Gorai River Restoration Project (GRRP) and the Ganges-Kobadak (GK) Project, both of which have significantly transformed the ecological and hydrological characteristics of the GDA, as described in the next chapter. Moreover, the two projects have created much greater hardship for the majority of agricultural households in Chapra, but have helped elite interests to consolidate and extend their control over water and agricultural resources.

References

Ahmed, I. 2013. Politics for Business or Business for Politics, trans. http://www.samakal.net/print_edition/details.php?news=20&view=archiev&y=2013&m=04&d=28&action=main&menu_type=&option=single&news_id=342189&pub_no=1392&type=, accessed May 1, 2013.

Bandyopadhyay, J. and N. Ghosh 2009. Holistic Engineering and Hydro-Diplomacy in the Ganges-Brahmaputra-Meghna Basin. *Economic and Political Weekly* 44(45): 50–60.

Bangladesh Bureau of Statistics (BBS) 2010. *Area Population Household and House-hold Characteristics*. Dhaka: Government of Bangladesh.

Bangladesh Water Development Board 2012. *The Ganges Basin Flow Data at Harding Bridge Point in Bangladesh*. Dhaka: Government of Bangladesh.

Bhattarai, D. P. 2009. An Analysis of Transboundary Water Resources: A Case Study of River Brahmaputra. *Journal of the Institute of Engineering* 7(1): 1–7.

Brichieri-Colombi, S. and R. W. Bradnock 2003. Geopolitics, Water and Development in South Asia: Cooperative Development in the Ganges-Brahmaputra Delta. *Royal Geographic Society with IBG* 169(1): 43–64.

Custers, P. 1993. Bangladesh's Flood Action Plan: A Critique. *Economic and Political Weekly* 28(29&30): 17–24.

Faber, D. and D. McCarthy 2003. Neo-liberalism, Globalization and the Struggle for Ecological Democracy: Linking Sustainability and Environmental Justice. In *Just Sustainabilities: Development in an Unequal World*. J. Agyemen, R. D. Bullard and B. Evans eds. Pp. 38–64. Cambridge, MA: MIT Press.

Government of Bangladesh (GoB) 1996. Treaty between the Government of the People's Republic of Bangladesh and the Government of the Republic of India on Sharing of the Ganga/Ganges Waters at Farakka. http://www.jrcb.gov.bd/attachment/Gganges_Water_Sharing_treaty,1996.pdf, accessed January 12, 2014.

Gupta, A. D., M. S. Babel, X. Albert and O. Mark 2005. Water Sector of Bangladesh in the Context of Integrated Water Resources Management: A Review. *International Journal of Water Resources Development* 21(2): 385–398.

Haftendorn, H. 2000. Water and International Conflict. *Third World Quarterly* 21(1): 51–68.

Islam, S. N. and A. Gnauck 2011. *Water Shortage in the Gorai River Basin and Damage of Mangrove Wetland Ecosystems in Sundarbans, Bangladesh*. Paper Presented at the 3rd International Conference on Water & Food Management (ICWFM-2011), Dhaka, January 8–10.

Iyer, R. R. 1997. The Indo-Bangladesh Ganga Waters Dispute. *South Asian Survey* 4(1): 129–144.

Khalequzzaman, M. 1994. Recent Floods in Bangladesh: Possible Causes and Solutions. *Natural Hazards* 9: 65–80.

Khan, T. R. 1996. Managing and Sharing of the Ganges. *Natural Resources Journal* 36: 456–479.

Mallet, V. 2014. Fears for Bangladeshi Democracy Rumble across Region. http://www.ft.com/cms/s/0/063cbe1a-79e5-11e3-a3e6-00144feabdc0.html#axzz35zyXCPQm, accessed January 13, 2014.

McGregor, J. 2000. *The Internationalization of Disputes over Waters: The Case of Bangladesh and India*. Paper Presented at the Australasian Political Studies Association Conference, Canberra, October 3–6. http://apsa2000.anu.edu.au/confpapers/mcgregor.rtf, accessed May 3, 2012.

Ministry of Water Resources (MoWR) 2001. *National Water Management Plan*. Dhaka: Government of Bangladesh.

——— n.d. Catchment Areas of Major Rivers. http://jrcb.gov.bd/bangla/index.php/9-link-page/12-basin-map. Accessed January 9, 2014

Nuruzzaman, M. 2007. Neoliberal Economic Policies, the Rich and the Poor in Bangladesh. *Journal of Contemporary Asia* 34(1): 33–54.

Padmanabhan, A. 2014. Teesta Water Pact a Difficult One: PM. http://www.newindianexpress.com/nation/Teesta-Water-Pact-a-Difficult-one-PM/2014/03/04/article2090140.ece, accessed June 13, 2014.

PTI 2014. Teesta Treaty Difficult Issue: PM Manmohan Singh Tells Bangladesh's Sheikh Hasina. http://www.dnaindia.com/india/report-teesta-treaty-difficult-issue-manmohan-singh-tells-sheikh-hasina-1966754, accessed April 5, 2014.

Rahaman, M. M. 2009. Integrated Ganges Basin Management: Conflict and Hope for Regional Development. *Water Policy* 11(2): 168–190.

Swain, A. 1996. Displacing the Conflict: Environmental Destruction in Bangladesh and Ethnic Conflict in India. *Journal of Peace Research* 33(2): 189–204.

Turton, A. and R. Henwood eds. 2002. *Hydropolitics in the Developing World: A Southern African Perspective.* Pretoria: AWIRU, University of Pretoria.

Water Resources Planning Organization (WARPO) 2001. *National Water Policy (NWP) 1999. Government of Bangladesh.* Dhaka: WARPO. http://www.partnersvoorwater.nl/wp-content/uploads/2011/07/WARPO2004_National-Water-Policy.pdf, accessed 2, 2013.

Zeitoun, M. and J. Warner 2006. Hydro-Hegemony-A Framework for Analysis of Trans-boundary Water Conflict. *Water Policy* 8: 435–460.

3 The Gorai River restoration and the Ganges-Kobodak projects

The government of Bangladesh has attempted to mitigate the negative effects of the Farakka Barrage by implementing projects intended to distribute the available water in more efficient ways. As documented in chapter two, immediately after the construction of this barrage in India, the flow of water during the dry season lessened by 70 percent in Bangladesh and never returned to its previous level. The flow of water during the rainy season also became much more unpredictable with significantly less water most years but with extreme flooding in others. These effects have been magnified in the case of the Gorai River with disastrous consequences for local farmers. In this chapter, I describe two major government projects, the Gorai River Restoration Project (GRRP) and the Ganges-Kobodak (GK) Project, both examples of "high modernism," as that term is used by Scott (1998). According to Scott, high modernism as an ideology creates legitimacy for scientific technologies, justifying, for instance, the construction of massive water infrastructures and centralized transportation and communication systems but without consideration for the voices of the poor and marginalized. In Bangladesh, I argue that both of the projects failed to take local ecosystems into account, and, therefore, they have worsened rather than improved the living conditions of the majority of farming households. I attribute these failures to the top-down approach of the central government, which gives local elites full control of the projects while excluding the majority of farming households from meaningful engagement in decision-making processes.

The Gorai River Restoration Project

The Bangladesh Water Development Board (BWDB) implemented the Gorai River Restoration Project (GRRP) since 1998, with the assistance of the World Bank, to help restore normal flows during both borsha and khora seasons by dredging the river channel. The reduced flow of water during the khora season causes an increase in sedimentation in the Gorai River (Islam et al. 2001; Mirza 1998). This sedimentation is called *charlands* when it acquires relatively permanent form as extended shoreline or islands in the river (Chowdhury 1984; Lahiri-Dutt and Samanta 2013). The increasing level of charlands further reduces river flow and causes ecological service failures. Ninety-one percent of the households I surveyed at Chapra, reported major challenges for practicing traditional agricultural activities due to failures of water supply and ecological services from the Gorai River and local water bodies. These water bodies include Chapaigachi oxbow lakes, Shinda and Shaota canals and the Lahineepara and Chapra wetlands, which were normally replenished each year by the Gorai River during the borsha season but sedimentation and erosion patterns now interfere with this process. The GRRP is a major initiative to reduce most of these negative effects.

This project is also a major example of corporate as well as governmental control over local water resource management in Bangladesh. It was implemented as part of the World Bank's Flood Action Plan (Boyce 1990; Paul 1995; Thompson and Sultana 1996) and Integrated Water Resource Management program in Bangladesh (Ahmad 2003; Brammer 1990; CEGIS 2003; Gupta et al. 2005). Phase one began in 1998 and continued until 2009. The dredging started at the point where the Gorai River branches originated in the Ganges at Kushtia and continued 20 kilometers downstream, including the portion of the river that flows past Chapra (de Groot and Groen 2001). Four foreign companies carried out this dredging at a total cost of US$160 million (Khan 2012). Costs were covered by several external donors, including the Dutch government and the World Bank (Ministry of Water Resources 2001; World Bank 1998). Phase two of the project began in 2010 and continued to 2013 with a target area of 36 kilometers from Kushtia city point of the river to Kumarkhali Sub-district. Phase two overlapped somewhat with phase one since new sedimentation and charland emerged in some phase-one areas after only few months due to the continuous basin flow reduction. The China Harbour Engineering Company (CHEC) executed the project under the supervision of BWDB. The total cost of the project was US$220 million; US$180 million was provided by the International Development Association and US$40 million by the national government of Bangladesh (World Bank 1998).

Implementation of the GRRP required the practices of a complex bureaucratic network. The Ministry of Water Resources (MoWR) coordinates and monitors the GRRP and, under the direction of the Prime Minister of Bangladesh, it established a GRRP national implementation committee. The committee is comprised of a Project Director from the Bangladesh Water Development Board (BWDB) and some other members from the departments of fisheries, agriculture, environment, local government, inland navigation, planning, finance, economic relations and the Joint River Commission. Some local government departments at the upazila and sub-district levels, such as the Local Governments and Engineering Department (LGED), in coordination with Non-Government Organizations and private sector representatives, executed some other tasks related to local water bodies like canal or wetland. Theoretically, based on the GRRP guidelines, the LGED is responsible, for instance, to restore linkages between the Gorai River and local water bodies in order to foster connections between water, crop lands and communities. Renovations of these water bodies are extremely important for restoring connection between croplands and the Ganges River. In addition to these LGED activities, the MoWR also formed a resettlement committee in coordination with NGOs to rehabilitate the project affected people. NGOs get involved in this project in support of the Poverty Reduction Strategy Paper and Millennium Development Goals (International Monetary Fund 2013; Matin and Taher 2001). Most of the project, however, is carried out by engineering corporations under the contract to one or another government agency. In this way, the central government is able to support corporate control as well as top-down management of local resources.

The corporate and foreign aid components of the GRRP are closely linked and as I argue in the sections that follow, the foreign aid and local development is best understood as "a form of eco-imperialism and as a 'debt trap'" as articulated by Wood (1984:703), and as perpetuating a vicious cycle of underdevelopment (Frank 1998; Ludden 2001:207; Prebisch 1963; Toye 1987:5; Wallerstein 2004:17–18). The central government imports foreign dredging technologies based on economic liberalization policies of the World Bank and International Monetary Fund (Baechler 1998:27; International Monetary Fund 2013; Nuruzzaman 2004) but, rather than realizing economic or social benefits, local populations experience mainly negative outcomes.

Elite domination at Chapra

The GRRP is a perfect example of the ways in which the political culture of Bangladesh and its top-down political processes tend to undermine local water management systems and promote corporate and elite control over local natural resources. The project was designed and implemented by the World Bank and the national government and, even when local government is involved, it is dominated by local elites. Local governments were responsible, for instance, to renovate local water bodies and connect them with the dredged area of Gorai River. Local governments were also responsible for ensuring proper working conditions and ensuring the security of dredging machines and staff. Many of the staff are foreigners and know nothing about local culture and tradition. They need a secure environment, food and lodging arrangements from local governments and, in order to service these needs, the dredging company works closely with local government leaders but not with the communities those leaders are supposed to represent. Local political leaders are thus able to direct the project in ways that favor their interests but do not necessarily serve the interests of other community members.

Fajal, Keramot, Laily and Harun, four focus group respondents, stated that local government officials make businesses from the GRRP development activities. In some cases the central government's elites and their close circle of political allies, also gain personal economic benefits from this project. One major example is the Padma Bridge corruption allegation of the World Bank (T. J. 2013). The World Bank alleged that high-level government staff and their relatives had accepted bribes from the international construction companies in exchange for the contract order. During the 2009 to 2013 project period, the illegal incomes of a nephew of a major political leader and the National Parliament's Deputy Speaker were reported to have increased by 32,985 and 4,435 percent respectively (*Prothom Alo* 2013). After the business interests of the central government are fulfilled, local governments and their associates in the local power structure exploit the project to fulfill their personal interests (e.g. *Daily Star* 2013). Local elites are the rich farmers and many of them hold local government positions such as chair or member of the Union Council (UC) or Upazila Parishad (UP). They control these local power structures and political party organizations at the district, sub-district, union and village levels. Based on these power structures, they control the GRRP implementation activities. Laily termed these activities as *"jor jar mulluk tar"* (powerful people control a constituency), which is helpful for protecting elite interests and for excluding marginalized community interests.

Elite domination causes major challenges for ensuring transparency and accountability. This domination creates an informal nexus between the UC and UP officials, bureaucrats and rich farmers. Focus group respondent Sabbir informed me how the official government plan stipulates that dredging should occur at a specific place but local elites often divert the dredging activities to another site. They do so in order to protect their own properties including cropland, fallow land and household infrastructures from river bank erosion and embankment failures. Other focus group respondents, Harun, Rahim, Fajal and Jobbar, argued that local elites deposit the GRRP dredged sand in places that make it easier to sell for commercial purposes. Dredged sand is very valuable in the construction industry where it is used to construct new buildings, roads, houses and bridges. Furthermore, they develop commercial activities like fisheries projects based on the dredging project.

Moreover, the GRRP illustrates a major gap between the government water management systems and local community needs and desires. According to the stated goals of the GRRP, the central government wants to develop flood control and facilitate drainage during the rainy season and secure enough water during the summer so that fresh water can flow to more southern points in the GDA including Khulna city, Mongla port and the Sundarbans. On the other hand, community members like Billal and Joardar want natural resources like siltation, algae, earth-worms and water hyacinth, back to promote cropland fertility. They also want to get back water-borne wild vegetables like marsh herb, water lily, ferns and hyacinth bean and natural fisheries like carp, barb and minnow in local water bodies. They want to be able to make agricultural materials like ploughshares, frames, ladders and sticks from natural resources like bamboo, cane and wood which is now in short supply. Natural resources like local wild vegetation are the major sources for nurturing domestic animals like cattle, chickens, ducks and goats. However, the GRRP did not recognize any of these community issues and concerns in the project proposal formulation and implementation stages that create major frustration in Chapra villagers.

Kusum, Rahim, Bimal and Faruk mentioned that the GRRP causes further challenges for existing usages. Chapra residents cannot take baths or satisfy their domestic water needs from the Gorai River. Kusum pointed out that it is difficult to predict the depth of the river and this leads to drowning deaths. During one visit to the region, on April 22, 2012, I saw a dead body floating in the Gorai River. Local people later informed me that this person had drowned when he was taking a bath. Again two students of Kushtia Medical College died and three others went missing when they took baths in a dredged area of the river on July 27, 2012 (*New Age* 2012a).

Tofajjal and Joardar argued that the GRRP project planners did not properly assess the river flow patterns and did not listen to those in the community who could have helped them understand those patterns. Alamgir, Bakar, Borkat and Mofiz stated that in one instance they explained to the contractor that the dredged sand should be deposited on the right bank of the river when he was depositing it on the left bank. Their logic was that the sand deposition on the right side could make the river bank stronger and reduce erosion and thereby protect buildings and cropland. On the other hand, the left bank is mostly charlands and does not require additional fortification. When the dredged sand is deposited on the left bank, it gets stronger and diverts the water more forcefully to the right side where erosion is already threatening buildings and fields.

Joardar, Billal, Faruk and Suman expressed feelings of intense frustration with these sand deposition issues because of their increasing concerns about the possible loss of household assets and croplands. They have shared these concerns with neighbors and submitted complaints to local government officials at the union and upazila levels. However, they are not hopeful about getting proper recognition of their voices from the government because they are not a part of the local power structure. Thus, they communicate with Tanvir, a wealthy landowner, who is a local leader of the ruling government. According to Joardar, Tanvir did not place enough importance on their concerns. Tanvir agreed with them about the negative effects of GRRP sand deposition. However, he argued that he did not have enough power to change these activities. Joardar explained Tanvir's intention differently. According to him, Tanvir is successful in gaining profits from the dredged sand. He did not want to lose these profits by raising local voices against the sand deposition.

Finding no other alternatives, local people at Chapra including Joardar, Alamgir, Harun and Kusum, demonstrated against the GRRP in front of the local government offices at the Kumarkhali Sub-district and Kushtia District levels. They composed different slogans against the ruling government due to their failures in protecting their local infrastructures and croplands. Because of this protest, local governments face the central government's dissatisfaction. The officials misinterpreted the protestors as opposition political party supporters who are unwilling to acknowledge the government's success in local water resource development. Therefore, local governments in coordination with local elites, like Tanvir, find ways to resist the GRRP protestors. Local governments impose sanctions against demonstrations, protests and even meetings at outdoor locations. According to many of my informants, they also arrested many protestors and perform extra-judicial killings of protest leaders.

Local elites also create financial hardships for protestors and threaten them physically. They raise demonstrators' debt issues with local banks and NGOs and encourage them to demand repayment of loan. Many of the local elites cancel sharecropping arrangements with people known to be among the protestors. They also have their supporters and hired musclemen to threaten protestors with killings, kidnappings and robberies. Rahim informed me that the musclemen do not hesitate to kill the front line demonstrators almost every year at Chapra. The governments support these illegal activities of local elites with every possible mechanism. Consequently, the GRRP negative effects are increasing survival concerns among the marginalized households at Chapra.

Sedimentation and charland

Despite the GRRP implementation, sedimentation and charlands are increasing concerns for Chapra residents. The head of the Gorai River, where it branches off from the Ganges, tends to become blocked by sedimentation as a result of the reduced flow in the Ganges River. Joardar informed me that charland creation had not been a dominant issue before the Farakka Barrage was constructed. Previously sedimentation was mainly considered helpful for cropland fertility and agricultural production. However, sedimentation has now become a curse because of the effect of charlands on river bank erosion. Jobbar, Morjina and Mofiz informed me that the increasing charland creates major challenges for agricultural production and traditional employment practices. Jalil and Raju emphasized the greatly increased scale of charlands production since construction of the barrage.

Charlands now inhibit the river flow throughout its course and the river runs in the different channels depending on the level of water. The Gorai River has been almost covered with these charlands, both in the middle of the river and along its banks, and the GRRP has failed to remove the majority of them. These charlands are not suitable for crop production because there is no stability in soil structure and water flow due to the irregular seasonal flows, storms or flooding. They are subject to encroachment of local elites and governments, however, which further deteriorates local ecosystems. Many local elites control the Gorai bank lands and build development infrastructures on them. They build fisheries projects, housing infrastructure, factories and business farms along the river banks. Local government agencies such as the Local Government and Engineering Department (LGED) worsen the problem by constructing roads and railways that block adjacent water bodies such as the Kaligangya River at Chapra. This river originates in the Gorai River and connects with local wetlands like the Chapra Beel. Construction of infrastructure for the GK project, which is described below, has also blocked many local water bodies including the Kaligangya, Sagarkhali and Dakua Rivers which creates major water stagnation problems at Chapra and neighboring areas. Sixty-two percent of my household survey respondents stated that they have encountered major water stagnation problems because of these development infrastructures. The stagnant water contributes to agricultural production failures, environmental pollution and water-borne diseases. Excessive rainfall or river flow increases the risk of these vulnerabilities. Joardar argued that this experience is different from the past. Their past experiences were that the river flow and rainfall meant greater natural resources like siltation, fisheries and water-borne wild vegetables.

River bank erosion and embankment failures

GRRP dredging has created more environmental vulnerabilities like extreme flooding or river bank erosion rather than resolving the existing problems. Joardar informed me that "*bonna, khora, nodi bhangong, ebong jolaboddota holo Allahor gojob, manuj jokhon prokriteke niyontron korar chesta kore prokrite ei gojob dei*" (when people dominate nature with technologies, nature gets angry and takes revenge by creating multiple flooding, drought, river bank erosion and water stagnation). Borkat, Fazlu, Sabbir and Kusum, also argued that the Farakka Barrage was the beginning of this *natural* punishment that is now made worse by the GRRP dredging, causing destruction of local infrastructures like houses, croplands and water bodies. The GRRP dredged sand is supposed to be deposited on the river bank so that the banks get stronger and help to reduce flood concerns. The embankments that have been built actually fail to control flooding; rather, they have increased the occurrence of sudden embankment failure and water stagnation.

According to Billal, "*odhik baads maane odhik moron faad*" (the increasing embankments mean increasing death traps). In the minds of many local farmers, the increasing technological domination over the Ganges Basin ecosystems causes nature to become exhausted and increases the amount and scale of vulnerabilities. My research clearly documents the fact that the Flood Action Plan and Integrated Water Resource Management in the GDA water management area have not restored the Gorai River flow and associated environmental characteristics and has not assisted agricultural productivity for the marginalized households at Chapra. The central government established the GRRP to flush out sedimentation and charlands from the Gorai River but, according to my informants, the project increased flooding, drought, river bank erosion, water stagnation and embankment failures. The construction of embankments to reduce river bank erosion has simply displaced erosion effects to other locations. Focus group participant Tarun points out that, new embankments divide and block local water bodies like the Kaligangya River at Chapra, a distributary of the Gorai, and create water stagnation and environmental pollution.

My focus group respondents also argued that the river bank erosion due to the GRRP is creating major, additional livelihood challenges. Joardar argued that this project destabilizes the ecosystem relationships between the Gorai River flows, river bank, vegetation and cropland areas at Chapra. Keramot, Jobbar, Suman and Laily, argued that, previously, the Gorai River banks were covered by wild vegetation, forest land, bushes and trees and stable soil textures and this supported stability of local housing patterns, water bodies and cropland use. Keramot and Jobbar argued that the larger scales of dredging produce larger embankment failures and greater areas of river bank erosion. These erosion and embankment failures increase with seasonal flooding or rainfalls that wash away the GRRP dredged sand. My household survey data identified that 76 percent of people at Chapra encountered river bank erosion in their lifetime. During my participant observation in 2011–12, I saw many rural roads, houses, trees and gardens partially inside the Gorai River due to this erosion. Forty-four percent of the household survey respondents reported encountering this erosion more than three times in their life. In Bangladesh as a whole, the total number of affected people because of this river bank erosion is 20 million who encountered losses of over US$462 billion over the last 36 years. Among the affected people, in south-western Bangladesh more than 10 million are now homeless (*Prothom Alo* 2014). The current top-down water management programs are a major reason for these disastrous consequences.

The GRRP destabilizes the river bank due to the continuous flow reduction caused by the Farakka diversion. This project can only achieve success if the Gorai River gets the regular water flow from the Ganges River. According to my household survey data, 89 percent of people at Chapra believe that the Farakka diversion is the root cause of this Gorai River bank erosion. According to Joardar, the erosion began immediately after resuming the Farakka Barrage in 1975 and increased further with the other technological interventions, like the Lalon Shah Bridge over the Gorai River. Joardar argued that the Ganges River has been turned away from Pabna district toward Kushtia district for seven kilometers after eroding the villages of Mohanagartek and Belkaloya. Every year, the Ganges Basin flow reduction causes sedimentation in the Gorai River mouth which blocks the Gorai water flow. The level of sedimentation and charland is increasing every year, and therefore, the river bank erosion is continuously dominating concerns over Chapra villagers. The GRRP cannot effectively remove the larger amount of sedimentation, which increases river bank erosion and embankment failures at Chapra.

Due to the GRRP dredging activities, the area downstream of the Gorai railway bridge at Koya is currently encountering river bank erosion. Again, the Gorai River right bank is eroding at Chapra. Khoksa Upazila, downstream of Chapra in Kumarkhali, has also faced this erosion for several years. The Gorai bank erosion at Khoksa causes major risks to the 500-year-old *Hindu Puja Mandir* or prayer house and *Hindu* crematorium. The village of Kamalpur, downstream of the Khoksa, is also encountering this erosion. Two-thirds of Hizlabot village has almost been destroyed due to erosion. Ganeshpur, downstream of Hizlabot village, is also encountering this erosion, which also destroyed areas further downstream, such as, Khoksa Upazila in Kushtia District, Shailokopa Upazila in Jhenaidah District and Sreepur Upazila in Magura District (*Daily Ittefaq* 2011).

Keramot, Harun, Tarun and Kusum argued that the river bank erosion and displacements are major causes of losing *bongshio shikor* or the 'ancestral roots" of farming households at Chapra. In 2011, for example, Chapra communities encountered embankment failure nightmares. Because of their concerns about embankment failures they spent several nights sitting up without sleep and stockpiled emergency goods like dry food and clothes, in case they had to abandon their residences. Billal bought fuel by reducing family food supplies to ensure available lights at a night to resolve potential emergency challenges like snake bites. Many of them visited the embankment site every 30 minutes to get updates on the embankment condition. Their fears were based on their past experiences of Gorai bank erosion.

In the context of socioecological dynamics, river bank erosion is a historical phenomenon at Chapra. For instance, Kafil, a focus group respondent, told me that his family has encountered serious erosion problems five times over a period of three generations. Many of his grandfather's croplands are now in the middle of Gorai River. He argued that the previous Chapra village is now charlands and the village has shifted two kilometers west in relation to the previous Gorai River channel. The current river location was their village during his father's generation. Erosion and embankment failures have significantly intensified, however, with the GRRP. Again, Billal informed me that he experienced this erosion from the very beginning of his childhood, as did his father, grandfather and great grandfather. His grandfather was a rich farmer, although he is currently a landless day laborer. Many croplands of his previous generations are now charlands and inside the Gorai River. His current residence is 20 feet distant from the river and is on the verge of erosion. The erosion also destroyed many local structures like hospitals, mosques, gardens and residences. The GK project infrastructure at Chapra, which I describe below, is now also at major risk of erosion and portions have already been destroyed twice.

Responses to my household survey data indicated that river bank erosion was the major cause of displacements for 86 percent of people at Chapra and, among them, 38 percent encountered this displacement more than three times. Laily and Kusum stated: *"nodivangon amader sob shesh kore disse"* (river bank erosion destroys our belongings). Every displacement disorganized their rhythm of livelihood, including household structures, cropland and agricultural production materials. Tanvir and Tofajjal faced major challenges for protecting their livelihood practices like agriculture, fishing, education, housing and health care. Again, Joardar and Billal lost their last household assets due to this erosion and displacement. Many displaced people left Chapra forever and now live in urban slums. Joardar informed me that a victim of the Rana Plaza garment collapse in 2013 was displaced from Chapra in 2007. The major reasons for this displacement are river bank erosion and flooding.

Flood vulnerabilities

Bangladesh is the third most flood vulnerable country of the world (UNDP 2004), and Chapra residents are among the most vulnerable in the country. My focus group and case study respondents express grievances against the GRRP based on the flood of 2011. Kafil asked me a basic question because of his current livelihood sufferings: "*aponi ki bolte paren amaderke ke rokka korbe ei durjog theke?*" (Could you please tell who will save us from disasters like flooding or drought?). He said that the Ganges River does not have enough water flow when we require it during the summer season. On the other hand, we encounter flooding when we need a regular flow. According to Joardar, the root cause of the 2011 flood was a sudden increase in flow due to combined effects of local rain and the river flow increase. In 2007, a second flood hit in October, when Joardar was still trying to recover from an earlier flood in August. The Gorai River lost its capability for transporting the sudden heavy water flow due to the major sedimentation and *charland* problems. When I was visiting their croplands at the Gorai River bank in 2011, Billal and Kafil showed me the aggressive flow of the Gorai River, and they termed this aggressiveness a *rakkhoshi nodi* or "monster river." According to Fajal, Keramot and Laily the GRRP is responsible for increasing flooding due to the increasing number and scale of embankment failures and river bank erosion. This flooding is even more dangerous than a regular flooding. In a normal flooding situation, local people can predict flooding characteristics and thus have time to save some household goods and assets.

Billal, Joardar, Jobbar, Suma and Morjina encountered the different types of flooding and their negative effects on household assets and field crops. Billal informed me that flood water from sudden embankment failures reaches to a rooftop within a moment and offers no scope for saving household assets, domestic animals, reserved food, local seeds and crops. Joardar encountered losses of agricultural crops, damages to his cooking place and losses of next season's seeds and production capital due to the flood in 2011. They lost domestic animals and shelter places and many of their household goods washed away in the flood water. Many of them encountered sickness, fevers and diarrhea. They lost household goods like utensils and structures like tube-wells. Community utilities like schools, electricity stations and mosques also encountered damage and loss. Billal, Joardar and Suman passed the rainy season enduring these types of socioecological vulnerabilities and started the winter season with new challenges. They had no crop seeds, agricultural production capital, bullocks or plows for resuming agricultural production. Moreover, the marginalized people like Joardar and Billal did not have money to spend on food, education, health care and housing. Many of them are desperate to find out possible borrowing sources but face exploitation by local elites and NGOs.

Chapra neighborhoods sought formal help from local government to recover from flood effects on river bank erosion in 2011 flood. However, they failed to get any positive response from local government, and therefore, they made traditional arrangements to overcome the embankment failure effects. They collected *harichaada* (a traditional system for contributing goods to a community event) from local people to raise money so that they can protect the Gorai embankment. Some community members contributed to this fund by cutting off daily expenditures on food and gas and selling domestic animals like chickens or ducks. They bought materials like bamboo, sand bags or poly bags with this money to make a stronger foundation for the embankment. Bilal argued that physical existence was the first priority (compared to starvation). Many community members provided manual labor, sacrificing time they could have spent at paid work. They put guards on the embankment site to get regular updates and ensure the embankment security. To make matters worse, after experiencing flood concerns during the rainy season, they encountered drought during the summer season.

Drought concerns

The GRRP often fails to reach the minimum objectives of the summer seasonal water flow for the basin communities at Chapra. Because of the Farakka diversion, the Gorai River discharge was reduced from 171 cubic feet per second (cusecs) in 1971 to 2 cusecs in 2003 (Islam and Gnauck 2011). Currently, the Gorai River is itself encountering survival challenges due to the longer drought season (Islam and Karim 2005). Local people cross the river on foot or use mechanized transportation because the river becomes so shallow during the summer season. While conducting my fieldwork at Chapra on March 2, 2012, I played football in the middle of Gorai River with community people. The Sundarbans mangrove forest further south is also encountering survival challenges and mangrove trees and marsh crocodiles are declining significantly due to the failures of the Gorai River based ecosystem in the Ganges Dependent area in Bangladesh (Mirza and Sarker 2004).

Fazlu, Bimal and Suman told us that the GRRP fails in restoring local water bodies due to the continuous Ganges Basin flow reduction. They stated further: *"ei desher dui vaag jol hoa sotteo pani aste aste durlov ebong beybohul hoyea jasse"* (Despite two-thirds of Bangladesh being covered with water earlier, it is getting scarce and expensive). Local water bodies like canals, water depressions or wetlands are drying up due to continuous failures of the Gorai River. Tofajjal informed me that local rivers like the Hisna, Kaligangya, Kumar, Hamkumra, Harihar and Chitra are disappearing from local maps due to the basin flow reduction. The disappearance of these rivers deteriorates surface and ground water levels. The GRRP fails in restoring these rivers and water levels. The disappearance of local water bodies is a major cause of failing river water supplies, crop land fertility, water-borne wild vegetables, fisheries and domestic animals. These failures are responsible for major agricultural production and employment challenges.

Kamal and Ibrahim, two focus group respondents, informed me that the GRRP fails in restoring the seasonal patterns of the basin flow. These failures increase challenges for agricultural production and employment opportunities. Joardar argued that sometimes the borsha or bonna ends early or later that alters the winter and summer seasonal patterns and this in turn changes the cropping patterns of kharif-2, robi and kharif-1 during the rainy, winter and summer seasons respectively. The khora season fails to get the regular water flow from the Ganges Basin due to borsha seasonal instabilities that create major challenges for robi and kharif-1 seasonal crop production. Focus group respondents reported that they encounter trauma every year during the rainy and summer seasons because of these instabilities. The GRRP fails in restoring the ecosystems. Currently, they no longer expect water and ecological services from the Gorai River for performing agricultural and livelihood practices.

The basin communities at Chapra are no longer hopeful about restoring the past agro-ecosystems on which they have relied on many generations. The reduced flow of water and the GRRP interventions disrupt these systems in the upper, mid-level and lower elevation croplands. The household survey data identified that 87 percent of respondents report facing major livelihood challenges as a result of recent changes. The cropping practices based on maacha and vita or raised platforms with bamboo and mud respectively are no longer helpful to produce crops. Many wild, freely available natural resources, described in chapter two, are also much less abundant under current conditions. Drought, in addition to flood and river bank erosion, is major reasons for this resource depletion, which are deteriorated by the continuous sedimentation and charland. The drought causes major challenges for the other water management projects like the GK in the GDA.

The Ganges-Kobodak project

The Ganges-Kobodak (GK) project, like the GRRP, is a state-owned water management project in Bangladesh. Since the 1970s its implementation has been based on the United Nations' Flood Control, Drainage and Irrigation (FCDI) Program (Alexander et al. 1998; Talukder and Shamsuddin 2012) and in support of the goals of the Green Revolution (Cleaver 1972:177; Herring 2001:235). Unlike the GRRP which is focused on the flow of water in the Gorai River, the GK project is focused on the construction and maintenance of a series of canals that divert water from the Ganges River at a point several kilometers upstream from where the Gorai River branches off from the Ganges. Also, unlike the GRRP, the GK project was originally implemented before the construction of the Farakka Barrage in 1955 with the purpose of providing a secure supply of irrigation water to farmers throughout the year. It was, in fact, the first large-scale irrigation project in Bangladesh. It was not completed until 1983, however, several years after the Farakka Barrage was constructed and, as a result, planners were forced to accommodate the changes brought about by the effects of Farakka diversion and mitigate them to whatever extent possible. As in the case of the GRRP, the GK Project has interrupted the natural movement of water in the Gorai River Basin, undermining the ecosystems on which most households depend and distributing the available water in ways that increase existing inequalities.

The project encompasses an area bounded on the north and east by the Gorai and Madhumati Rivers, on the south by the Nabaganga River and on the west side by the Mathabhanga River (Map 3.1). Due to the decreasing flow of water

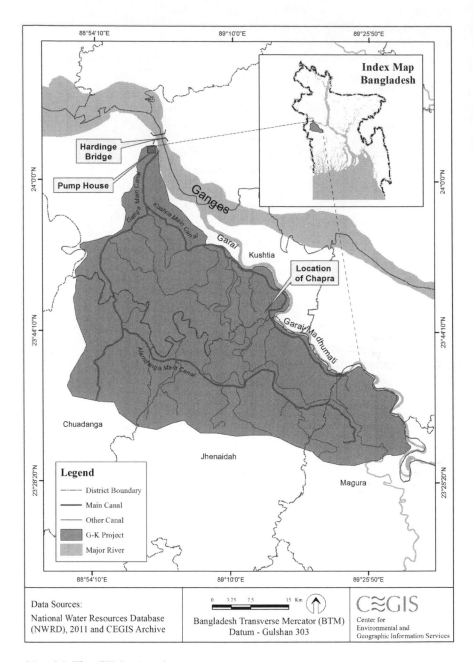

Map 3.1 The GK Project Area

in the Ganges and Gorai Rivers the GK canals are currently the main source of irrigation water for agricultural production in Chapra and throughout the GDA, despite the fact that the project had been established originally as a supplementary water supply.

The project has 2 main pumps with 3 subsidiary and 12 tertiary pump units (Table 3.1) capable of discharging 5,400 cusecs of water (Bangladesh Water Development Board n. d.). The project has an electricity requirement of 14 megawatts and has its own grid substation for providing this electricity. There is a 705-meter long intake channel that pulls the Ganges water to the main GK canal based on lift-cum-gravity flow for agricultural production. The main canal and sub-canals have 3,500 outlets to control irrigation water supplies in croplands. The project authority has established 2,184 hydraulic structures for the entire project area in addition to these outlets. These structures control water supplies at the main, secondary and tertiary canals. Project infrastructure also includes 228 kms of roads to promote project performance and resolve farmers' water demand concerns (Bangladesh Water Development Board n.d.).

There are three main canals of the GK Project that cover 193 kilometers (Table 3.1). The secondary canal system covers an additional 467 kilometers and the tertiary system covers another 995 kilometers, a total of 1,655 kilometers. Most of the main canal is made with concrete and the secondary and tertiary canals are almost completely made of packed earth. All of the main, secondary and tertiary canals are interconnected and provide a lift-cum-gravity method of water supply from the source water point on the Ganges River to local cropland. The project also includes 39 kilometers of flood control

Table 3.1 GK project infrastructure

Infrastructure	Sub-section	Quantity/Capacity
pump houses	main pump	2 units
	subsidiary pump	3 units
	tertiary pump	12 units
water-carrying capability	main pump house	3900 Cusec
	subsidiary pump	1500 Cusecs
total outlets		3500
inspection roads		228 kilometers
irrigation canals	main canals	193 kilometers
secondary canals	467 kilometers	
tertiary canals	995 kilometers	
flood control embankments		39 kilometers
drainage canals		971 kilometers

Source: Bangladesh Water Development Board n.d.

embankment. Additionally, the project has a 971-kilometer long drainage canal for transporting extra water from the project area to overcome water stagnation (Table 3.1). The project has also constructed 2,184 fish projects to provide a supplementary fish supply for local marginalized communities who are affected by the project infrastructures.

The project encompasses a total area of 488,032 acres of which 286,642 acres of cropland receive irrigation water (Table 3.2). It covers 13 upazilas within the four districts of Kushtia, Chuadanga, Jhenaidah and Magura. The total population of the GK project area is 2.5 million. Among this population, 150,000 household heads own enough agricultural land to obtain the project water service. These households represent 60 percent of all farming households in the area according to the Bangladesh Water Development Board (n.d.). Based on my fieldwork experiences, a person who has half an acre or more of cropland is capable of affording the project water supply for agricultural production. The remaining 40 percent do not have enough cropland to afford these services. Among the 60 percent 2 percent are rich households (3,000 households in total in the project area; 13 percent are intermediate (19,500 households) and 85 percent are small farmers (127,500 households) (Bangladesh Bureau of Statistics 2005).

The GK authority requires a massive organizational system for ensuring proper management of water supplies, infrastructural maintenance, accountability and transparency. The project has a complex chain of command from the central government to field supervisors. The central government established an office in Kushtia for operations and maintenance under the direct control of the BWDB head office in Dhaka. This local office operates under a Project Director who is at the rank of superintendent engineer overseeing several departments, for example, hydrology, morphology and agriculture. The departmental heads are also engineers. They supervise various sub-sections under their departments and coordinate field level activities with field staff in the project area. The field staff are responsible for collecting infrastructural and water updates to develop better performance of the project. The project's yearly operation and

Table 3.2 Cropland areas and population served by the GK project

Project Target	Target Description
project coverage area	488,032 acres
irrigation target area	286,642 acres
project area population	2.5 million
owners of land	150,000
location	Kushtia, Chuadanga, Jhenaidah and Magura

Source: Bangladesh Water Development Board n.d.

management costs are reported to be US$24,881,100[1] (Bangladesh Water Development Board n.d.).

In keeping with the emphasis on agricultural water use, the GK Project maintains a laboratory to test High Yielding Variety (HYV) seed quality to determine its suitability for crop production in the project area. The government also established a training center in Kushtia city for local farmers' skill development, so that they can farm the HYV crops based on the project water supplies. All of these activities are closely monitored by six major departments of the project office in Kushtia. The project also developed guidelines for community participation in local water management. For this purpose, there are 749 water management groups based on local farmers' participation. These groups are divided into 49 farmers' clubs, which are again divided into 7 water management associations. The water management federation is formed with these associations. However, as noted for the GRRP, this local water management system creates success for the rich farmers and failures for the marginalized farmers. The rich farmers control local water management associations as many of them are chairs or members of local Union Council or Upazila Parishad.

Outcomes of the GK project

The project has had some success in providing water supplies for agricultural production in Kushtia and Jessore Districts. For example, in 1968–69, it provided irrigation water for 67,933 and 14,384 acres of cropland to produce aman and aus rice paddy crops respectively in these two districts (Bangladesh Bureau of Statistics 1983). By 1978–79, in Kumarkhali Sub-district within Kushtia District, where Chapra is located, the GK project was successful in providing irrigation water to 14,287 and 4,451 acres of aman and aus cropland respectively (Table 3.3). In 1979–80, the water supply in Kumarkhali Sub-district was reduced to 10,685 acres of irrigated aman cropland during the rainy season but the water supply for aus during the summer season increased to 6,268 acres. The project faced serious challenges, however, when local water demand for robi crop production during the winter season increased when the source water

Table 3.3 GK project irrigated acreage at Kumarkhali Sub-district

Year	Aman (acres)	Robi (acres)	Aus (acres)
1978–79	14287	—	4451
1979–80	10685	—	6268
1980–81	12012	508	7929
1981–82	12594	122	8210
1982–83	13330	—	9354

Source: Bangladesh Bureau of Statistics (1983)

was decreasing due to the Farakka Barrage diversion. The GK Project thus had some successes but failed to provide a year around sources of irrigation water with seasonal dynamics (Table 3.3). The project also created a set of additional hardships for the majority of local farming households; (i) canal and road construction has caused significant damage to local ecosystems and associated natural resources; (ii) project infrastructure creates more vulnerability to both sudden floods and water stagnation; (iii) the project transforms water into a commodity; and, (iv) the project provides elites more control over local water resources.

Kabir, Kusum and Laily argued that the failures of the Gorai River management make them further dependent on the GK project for livelihood practices. Billal uses the project water for bathing domestic animals, e.g., cows and goats. Many of the respondents also take baths in this water. They raise domestic ducks and cattle based on this project water. Kusum and Laily collect water from the project to perform domestic activities like cooking and cleaning. Farm workers drink this project water when working in the crop fields. They also take rests under the GK project orchard. Many of them get seasonal fruits like mangos and jackfruits from areas irrigated by the project. They also cultivate seasonal vegetables on the project canal bank areas. Many of them harvest fish from the project water for household consumption. They are forced to use the GK project water for these purposes, due to the Gorai River water failures. The canals thus provide a few of the ecosystem services formerly available from the river, but these services are now available in a much diminished capacity.

Construction phase conflict and mismanagement

The GK project construction further promoted domination of the ruling elites in local water management. Tofajjal argued that the rich people who were inside the power circles supported the project's construction and the marginalized people were very concerned about losing cropland, houses, stored foods and production materials. As noted above, the GK project area has a population of 2.5 million and among them, only 150,000 households (Bangladesh Water Development Board n.d.) with a total population of about 700,000 are direct beneficiaries (based on average household size of 4–5 people). The remaining population of 1.8 million people includes urban and rural households, a large percentage of whom are marginalized, agriculture-dependent sharecroppers, day laborers, fishermen, boatmen, blacksmiths and other trades people, all with less than one-half an acre of land are excluded from the project's formal water services. Although exact figures are not available, it is clear that the majority of agriculture-dependent households are not direct recipients of the GK project services. Despite their inaccessibilities of the services, they have experienced damages and loss of household infrastructure and common property resources like fisheries as a result of canal construction. On the other hand, the rich farmers have been successful in gaining more benefits from the project construction. Tanvir's house was a center point of the project construction at Chapra. The project staff lived in his parents' house. Tanvir and his father considered these activities as social services although their close contacts with the project staff and control over project compensation money helped them to fulfill personal interests. Therefore, they were successful in saving assets like croplands, household structures and fisheries projects during the canal construction.

The GK project construction faced resistance from the local marginalized people at Chapra due to a number of concerns: the loss of cropland and household infrastructure, displacement, forced changes in cropping patterns and water commodification. My household survey data confirms that 56 percent of my respondents tried to stop the GK project. They failed, however, because of the rich farmers' non-cooperation and domination. Focus group respondents Kamal, Kusum, Alamgir and Morjina informed me that the government used the legal system to stop the protesters. Some local residents including Kamal submitted their complaints of eviction from residences and croplands to the court in Kushtia. However, their complaints were rejected because of the power structure's domination over the government system. Consequently, 43 percent of my survey respondents lost household assets and cropland. The project authority acquired these croplands and houses to build canals, flood control structures, roads and water bodies. Joardar lost his home and currently he lives alongside the project canal roadside at Chapra. Many displaced people lost their *bongshio shikor*, or "ancestral roots," at Chapra and now live in slums, on dams and embankments. Many of the displaced people moved to one of the 6,000 slums in Dhaka (*Daily Jugantar* 2014).

Most of the people at Chapra who own cropland first experienced water commodification for agricultural production with the GK project. This commodification creates extreme economic burdens for the small farmers who are, as I mentioned, 85 percent of the total project farmers. Another 41 percent people who fail to get the project services encounter further livelihood difficulties like food consumption, due to price hike because of production cost increases. Their access to safe water for drinking and other household activities also deteriorates as a result of this commodification. Rahim and Kafil argued that the Gorai River water and ecosystem services were free and readily available before construction of the Farakka diversion. However, the Gorai flow reduction makes them absolutely dependent on the GK project. Finding no other alternatives, they accept this water commodification and turn into water consumers, although many of their croplands and houses are within 20 meters of the Gorai River.

Water commodification displaces egalitarian water practices and creates an additional gap between the rich and marginalized people at Chapra. Their traditional water sources from the oxbow lakes, canals and wetlands that used to irrigate croplands, are no longer available due to failures of the Ganges flow and the Gorai River. As a result, their traditional irrigation practices like water pulling and drainage canal building with natural resources like bamboo, rope and mud are no longer useful. The rich farmers can overcome and adapt to these changes because of their surplus land ownership and access to capital. They have greater water demands and more influence within the GK management system. Their previous sense of community water ownership, responsibility towards community people, local maintenance systems and traditional irrigation practices no longer apply. Local ecological knowledge about the seasonal timing of water supplies and close observations of crop and soil health are no longer applicable based on the current practices. Thus, the GK Project has had the effect of breaking down traditional systems of reciprocity and mutual dependence between different classes, removing the dependence of the rich on the poor and further marginalizing poor households.

Operational phase conflict and mismanagement

Resistance to the GK Project was not limited to the construction phase but continued during the operational phase as well. The source point of the project at Bheramara is encountering severe sedimentation, charland deposition and river bank erosion every year which reduces project performance. The project authority performs maintenances regularly to keep the project source water point unblocked but the distance from the source point to the Ganges River is increasing due to continuous Ganges River flow reduction and sedimentation. Consequently, the majority of my survey respondents who are the project subscribers, failed to get a sufficient water supply from the project and 84 percent of my survey respondents believe that the Farakka diversion is the main reason for this failure. Their major complaint is that the Farakka diversion reduces the Ganges flow in Bangladesh, and therefore, the project fails to get sufficient water and encounters multiples problems.

Kamal, Ibrahim and Abul reported the uncertainty of the GK water supply makes it very difficult for them to maintain the cropping schedule of kharif-1, kharif-2 and robi, which are described as aus, aman and winter seasonal crops respectively. Kamal and Ibrahim further explained that the GK authority often provides water earlier or later than needed depending on water availability of the Ganges River. They encounter surplus water supply problems during the rainy season when they require less water and fail to get enough water during the summer season when they desperately need water supplies. Both surplus water and drought are responsible for their agricultural production damage and loss. Among my survey respondents, 78 percent reported that they failed to get responses from the GK staff to resolve these problems and 63 percent of them reported agricultural production loss as a result.

Jalil and Raju, focus group respondents, reported that they do not receive any GK water as their croplands are at distant places. The project provides water on a plot-to-plot basis using gravity flow from the Ganges River. Croplands close to the main canal get surplus water, whereas distant croplands fail to receive minimum water supplies. Joardar elaborated this problem in the different dimensions; the lower and higher croplands are subject to water stagnation and drought respectively.

Kafil, Joardar, Malek and Faruk have taken desperate steps to bring GK water to their croplands. Kafil removed top soil from the higher croplands and deposited it on the lower croplands to raise the platforms to overcome water stagnation. Joardar built an informal earthen-canal across the land of other farmers in order to transport the GK canal water to his distant cropland. Many farmers create temporary blockades across the project canals in order to move water to their land. Kafil sometimes uses temporary pipelines to get water and, after fulfilling his demands, he removes the pipelines.

Suman, Sabbir, Kusum and Fazlu report major losses of HYV crop production due to the GK project's irregular water supplies. This HYV crop production requires controlled water and therefore both water scarcity and extra water cause a loss of crop production. One major example of an over-supply of water occurred in 2011. Based on past water supply experiences, local farmers sowed IRRI (T-32) paddy, a HYV rice species, in mid-June and expected to harvest this crop at the beginning of August. However, they experienced a sudden rainfall on July 3, 2011 that damaged the IRRI (T-32) crops. The project drainage channel and flood control system failed to remove excess water from croplands which caused water stagnation and crop damages.

The failure of the GK project forces farmers to adopt alternative cropping strategies at Chapra. Rahim and Kafil report facing severe water crises almost every year during the kharif-1 season when they have their highest water demand. Kharif-1 occurs during the period from March to May based on the seasons of mid-spring and mid-summer. In this season, the project is only able to provide water supplies, on average, to 286,642 of 488,033 acres of croplands every year since its full operation (Bangladesh Water Development Board n.d.). Based on these past experiences, farmers produce alternative crops like lentils, beans, wheat and onion that do not require the project water. However, the GK project authority does not want to recognize these alternative cropping practices because this recognition could be interpreted as evidence of the GK project failure. This desperation causes major challenges for crop production of the marginalized farmers at Chapra.

In 2012, for example, the GK authority suddenly received surplus water from the Ganges flow during the robi cropping season although local farmers had sowed crops like pulses, mustards, wheat and onions, which cannot survive in wet soil. The robi cropping season is the period from November to February in late autumn, winter and early spring. My in-depth case respondents, Tanvir, Tofajjal, Joardar and Billal, requested the GK authority to stop the water supply because they were at the final stage of cultivating these crops. However, the authority ignored their request because they wanted to demonstrate that the project is successful in supplying water during the robi season. Joardar, Alamgir and Billal termed the GK project authority's act as *"paka dhane moi dea"* (destroy a success in agricultural production when it is almost ready to be enjoyed). Consequently, 10,000 acres of field crops were damaged and local farmers lost about US$1,491,554 (Khulna News 2012). These crops were essential to their livelihoods as they attempted to recover from the flood and drought of 2011.

As I discuss below, this destruction is a direct outcome of the lack of accountability of project managers to local people.

Lack of accountability and transparency

Joardar argued that the project authority did not bother much about marginalized farmers' water concerns. The project operates independently of local water management institutions and is directly controlled by central government agencies. As a result the GK local authority does not attach enough importance to local water concerns like water stagnation or drought. According to Jobbar and Fazlu, the GK field staff visit the rich farmers regularly but do not care about the water problems of the marginalized farmers.

Lack of accountability has also led to operational problems. Many canals have been out of order for years and many tertiary canals are disconnected from the main canal. Machinery of various kinds is stolen or inoperable. Consequently, rainwater spills into the canals and damages canal infrastructure. Kamal and Rahim informed me that many of the secondary and tertiary canals suffer blockages due to these maintenance problems. Portions of some drainage canals are covered with grass, bushes and rat holes that create water flow blockages and drainage congestion. These problems are further compounded by canal water leakage, misuse and wastage. Water leakage creates water scarcity for some croplands and surplus water for other croplands. Forty-four percent of my household survey respondents reported that they asked GK authorities to repair such problems but nothing was done.

Due to lack of transparency, project funds are mismanaged and this is also a major cause of declining project performance. Kafil and Ibrahim informed me that many conductors of the GK project do not perform their assigned tasks, like repairing the project area, although they are successful in getting the full contract money. No government agency raises questions to the project authority because they share benefits from these malpractices. Joardar informed me that a contractor received a contract for a 5 kilometer long repair job for the tertiary canal at Lahineepara in 2010. The contractor performed the assigned tasks only for half a kilometer although he took away the total money. Contractors have been doing these malpractices for a long time and people rarely complain to the local or central governments because of the lack of transparency and accountability in the government systems. Joardar argued that the project staff dismissed local complaints as politically biased activities against the government.

Kafil and Ibrahim explained to me the hardships they face when lodging complaints against the project malpractices. The Kushtia District project office is responsible for more than 800 kilometers of canal infrastructure. Kamal needs to spend a whole day to reach this office, if he wishes to submit a complaint. Despite this day-long effort, he is often not able to meet with the Project Director directly due to the complex bureaucratic hierarchy. In addition to the Project Director, the project has several department heads, assistants to heads and section officers. Joardar needs to respond to several questions at every stage of this hierarchy before he reaches a department head and hopefully the Project Director. Joardar does not understand this office culture and feels that the bureaucrats treat him unfairly. He also reports that the authority does not fully recognize the importance of his water concerns as officers behave like aristocrats and describe them with mental mapping as like slaves. As a result, he is not comfortable visiting this office and often does not report water problems.

The GK authority also fails to resolve local water conflicts between the rich and marginalized farmers. Many rich farmers have built facilities that allow them to reserve project water for later use on their croplands. When the other farmers request the release of this water, the rich farmers get angry and they begin to treat each other like enemies. Fifty-eight percent of my survey respondents at Chapra report having executed illegal actions to obtain the GK Project water under these circumstances. Many marginalized farmers cut off aisles of the rich farmers' croplands at night illegally. Many of them block a tertiary canal and divert the water to their land. Tanvir, a wealthy farmer, describes these actions as the outcome of hatred and jealousy against his or other wealthy farmers' success, and, consequently, he takes legal action against the marginalized farmers. On the other hand, Joardar describes these actions as revengeful activities based on Tanvir's control over the local power structures. The GK authority and local governments fail to recognize the marginalized people's frustration. Law enforcement agencies execute legal actions such as law suits, jails and fines and also harass innocent farmers in order to protect the interests of the wealthy (Islam 2013).

Joardar and Billal further informed me that the rich farmers initiate some informal actions. They withdraw sharecropping lands from the marginalized farmers who are not loyal to them. They also fire agricultural staff who do not support their water control over local resource management. Likewise, they exclude the marginalized people from the government safety-net programs like employment, housing, health care or education. The local elites put pressure on local banks and NGO authorities to get their loans back from these marginalized farmers. In some cases, they are able to ostracize individuals within their community so that they cannot walk to neighboring lands or on public roads to perform everyday practices like schooling, marketing or cultivation.

Even though GK project staff and contractors ignore many of the infrastructural problems associated with canal maintenance, they do respond positively to requests from the rich farmers. Joardar reported facing water shortages during the 2010 kharif-1 crop season and he made several requests to GK staff. However, the staff ignored his requests because of Tanvir's water concerns. If the staff had responded positively to Joardar's request, Tanvir's croplands would have encountered a water shortage. Because of this locally embedded power structure, Joardar encountered production loss and Tanvir was successful in ensuring his production. In every village, a Union Council member and his associates, like Tanvir, own a major portion of local croplands. If something happens against their interests, they exploit the local power structures to secure their control over the GK project.

The central government has been forced to establish some alternative sources of irrigation water due to the GK project's water supply failures. The government introduced shallow tube-well and deep tube-well construction programs for this purpose. More than one tube-well project was established for every *chalk,* or cluster of croplands, in an agro-ecological area. The rich farmer, Tanvir, was able to construct private tube-wells inside the GK project area and, as a result, he withdrew many of his croplands formally from the GK project. However, he left some croplands under the GK project in order to retain membership in the project water management association. He is, thus, still able to control the GK project water while gaining extra security through the use of tube-wells. The GK project in combination with government supported tube-well construction has thus reinforced elite control over local water resources rather than provided more equitable access.

The pattern described here is consistent with what Das and Poole have referred to as the "colonization of margins" (2004:3) and with the colonization of what Habermas (1987) calls the "lifeworld." The intrusion of the state in this setting follows a typical pattern, in which scientific knowledge and techno-centric water management policies are developed, consistent with neoliberal approaches to agricultural modernization and water commodification. This approach fails to recognize community voices and supports elite control over local resources (Agrawal 1995; Leach et al. 1999:225; Peet and Watts 1996b:11–16). Elite interests – from the local level at Chapra to the international scale at the World Bank – can claim this approach is successful at the same time as local marginalized people are encountering continuous ecosystem service and livelihood failures.

Conclusion

The top-down water development approach that is evident with the GRRP and the GK project fails to protect Chapra villagers from livelihood challenges. The GRRP fails to protect the Gorai River and local water bodies like Chapra *beel* that extend failures of ecological resources like siltation, water-borne vegetables and fisheries. In this context, there are no scopes for getting rainy, winter and summer seasons in coordination with *kharif*-2, *rabi*, and *kharif*-1 cropping patterns. The failures of Gorai River cause the major grounds of searching for alternative sources of irrigation services for agricultural production. The GK project is the major alternative irrigation source for this purpose. The project had some initial success in water supply but fails to continue this service due to the Ganges flow irregularities. The source water point of the project is going further away from the basin and encountering sedimentation and charlands. Moreover, the project fails to develop proper mechanisms of water supply in coordination with agro-ecological system. Again, the continuous ecological vulnerabilities of flood, drought, water stagnation, river bank erosion, and embankment failure reduce the GK project performances.

The GK project has failed to develop an effective or equitable water and agricultural management system. Both projects increase the inequality of rich and marginalized farmers. Both work in support of the central government's agenda of introducing HYV crops, chemical fertilizers and pesticides to replace local crops, siltation and algae and traditional plowing. Therefore, the projects transform agricultural production from traditional practices to market economic systems that induces the major gap between the rich and marginalized farmers. The rich farmers are successful in promoting commercial agricultural production and agri-business based on these technologies. On the other hand, the marginalized farmers are encountering livelihood transformations and survival challenges. In the next chapter, I focus in more detail on the transformations that have occurred to agricultural practices in the Chapra region as brought about by the three major government interventions I have described: the Farakka diversion, the GRRP and the GK project.

Note

1   This figure represents the conversion value of Bangladesh taka to US dollars as of May 16, 2012. US$1= BDT 80.3823.

References

Agrawal, A. 1995. Dismantling the Divide between Indigenous and Scientific Knowledge. *Development and Change* 26(3): 413–439.

Ahmad, Q. K. 2003. Towards Poverty Alleviation: The Water Sector Perspectives. *Water Resource Development* 19(2): 263–277.

Alexander, M. J., M. S. Rashid, S. D. Shamsuddin and M. S. Alam 1998. Flood Control, Drainage and Irrigation Projects in Bangladesh and Their Impacts on Soils: An Empirical Study. *Land Degradation & Development* 9(3): 233–246.

Baechler, G. 1998. Why Environmental Transformation Causes Violence: A Synthesis. *Environmental Change and Security Project Report* 4: 24–44.

Bangladesh Bureau of Statistics (BBS) 1983. *The Ganges-Kobodak (GK) Project Water Supply*. Dhaka: Government of Bangladesh.

——— 2005. *Key Information of Agriculture Sample Survey. Government of Bangladesh*. Dhaka: BBS.

Bangladesh Water Development Board N.d. *Ganges-Kobodak (GK) Irrigation Project*. Dhaka: Government of Bangladesh.

Boyce, J. K. 1990. Birth of a Mega Project: Political Economy of Flood Control in Bangladesh. *Environmental Management* 14(4): 419–428.

Brammer, H. 1990. Floods in Bangladesh: II. Flood Mitigation and Environmental Aspects. *The Geography Journal* 156(2): 158–165.

Center for Environmental and Geographic Services (CEGIS) 2003. *Analytical Framework for the Planning of Integrated Water Resource Management*. Dhaka: CEGIS.

Chowdhury, M. A. 1984. *Integrated Development of the Sundarbans*. Bangladesh: Silvicultural Aspects of the Sundarbans, FAO Report No/TCP/BGD/23 09Wf, W/R003.

Cleaver, H. M. 1972. The Contradictions of Green Revolution. *The American Economic Review* 62(1&2): 177–186.

Daily Ittefaq 2011. The River Bank Erosion on the Gorai River Created Flood in Five Districts, trans. http://www.ittefaq.com.bd/index.php?ref=MjBfMDFfMjFf MTRfMV8yNV8xXzEwMjYwMw==, accessed October 25, 2011.

Daily Jugantar 2014. Total number of Slums in Bangladesh is 13,930. http://www. jugantor.com/news/2014/04/27/92940, accessed June 3, 2014.

Daily Star 2013. Wealth, Income Soar in 4 Years. http://archive.thedailystar.net/ beta2/news/wealth-income-soar-in-4-years/, accessed June 15, 2013.

Das, V. and D. Poole eds. 2004. *Anthropology in the Margins of the State.* Santa Fe: School of American Research Press.

De Groot, J. K. and P. V. Groen 2001. The Gorai Re-Excavation Project. *Terra et Aqua* 85: 21–25.

Frank, A. G. 1998. *Reorient: Global Economy in the Asian Age.* Berkeley: University of California Press.

Gupta, A. D., M. S. Babel, X. Albert and O. Mark 2005. Water Sector of Bangladesh in the Context of Integrated Water Resources Management: A Review. *International Journal of Water Resources Development* 21(2): 385–398.

Habermas, J. 1987. *Theory of Communicative Action: Critique of Functionalist Reason.* Volume 2. Oxford: Polity Press.

Herring, R. J. 2001. Contesting the Great Transformation: Local Struggles with the Market in South India. In *Agrarian Studies: Synthetic Work at the Cutting Edge.* J. C. Scott and N. Bhat eds. Pp. 235–263. New Haven and London: The Yale ISPS Series.

International Monetary Fund (IMF) 2013. *Bangladesh: Poverty Reduction Strategy Paper (PRSP).* Washington: IMF.

Islam, G. M. and M. R. Karim 2005. Predicting Downstream Hydraulic Geometry of the Gorai River. *Journal of Civil Engineering* (IEB) 33(3): 55–63.

Islam, M. R., Y. Yamaguchi and K. Ogawa 2001. Suspended Sediment in the Ganges and Brahmaputra Rivers in Bangladesh: Observation from TM and AVHRR. *Hydrological Processes* 15(3): 493–509.

Islam, M. S. ed. 2013. *Human Rights and Governance Bangladesh.* Hong Kong: Asian Legal Resource Centre.

Islam, S. N. and A. Gnauck 2011. *Water Shortage in the Gorai River Basin and Damage of Mangrove Wetland Ecosystems in Sundarbans, Bangladesh.* Paper Presented at the 3rd International Conference on Water & Food Management (ICWFM-2011), Dhaka, January 8–10.

Khan, A. R. 2012. Irregularities Cloud Gorai River Dredging. http://arrkhan. blogspot.ca/2012/02/irregularities-cloud-gorai-river.html, accessed May 23, 2013.

Khulna News 2012. Magura Villages Lost 12 Crores with Crop Damages. http:// www.khulnanews.com/divisional-news/magura/12679-2012-01-25-05-26-28. html, accessed September 13, 2012.

Lahiri-Dutt, K. and G. Samanta 2013. *Dancing with the River: People and Life on the Chars of South Asia.* New Haven: Yale University Press.

Leach, M., R. Mearns and I. Scoones 1999. Environmental Entitlements: Dynamics and Institutes in Community-Based Natural Resource Management. *World Development* 27(2): 225–247.

Ludden, D. 2001. Subalterns and Others in the Agrarian History of South Asia. In *Agrarian Studies: Synthetic Work at the Cutting Edge*. J. C. Scott and N. Bhat eds. Pp. 205–235. Yale: The Yale ISPS Series.

Matin, N. and M. Taher 2001. The Changing Emphasis of Disasters in Bangladesh NGOs. *Disaster* 25(3): 227–239.

Ministry of Water Resources (MoWR) 2001. *National Water Management Plan*. Dhaka: Government of Bangladesh.

Mirza, M. M. Q. 1998. Diversion of the Ganges Water at Farakka and Its Effects on Salinity in Bangladesh. *Environmental Management* 22(5): 711–722.

Mirza, M. Q. and M. H. Sarker 2004. Effects on Water Salinity in Bangladesh. In *The Ganges Water Dispersion: Environmental Effects and Implications*. M. Monirul Qader Mirza ed. Pp. 81–102. Netherlands: Kluwer Academic Publishers.

New Age 2012a. 2 Students of Kushtia Medical College Drown. http://newagebd. com/detail.php?date=2012–07–28&nid=18706#.U7CMlO_QfIU, accessed September 25, 2012.

Nuruzzaman, M. 2004. Neoliberal Economic Policies, the Rich and the Poor in Bangladesh. *Journal of Contemporary Asia* 34(1): 33–54.

Paul, B. K. 1995. Farmer's Response to the Flood Action Plan (PAP) of Bangladesh: An Empirical Study. *World Development* 23(2): 299–309.

Peet, R. and M. Watts 1996b. Liberation Ecologies: Development, Sustainability, and Environment in an Age of Market Triumphalism. In *Liberation Ecologies: Environment, Development, Social Movements*. R. Peet and M. Watts eds. Pp. 1–23. New York: Routledge.

Prebisch, R. 1963. *Toward a Dynamic Development Policy for Latin America*. New York: UN Economic Commission for Latin America.

ProthomAlo 2013. We Do Not Know Khaleda, Do Not Go Near Hasina: We Do Not Want the Sick Government, trans. http://www.prothom-alo.com/bangladesh/ article/#Scene_2, accessed February 3, 2014.

——— 2014. No Government Preparation Is Available: Two Crore People Are Victims of River Bank Erosion for the Last 36 Years and One Thousand Losses Every Year. http://prothom-aloblog.com/posts/16/46251/, accessed June 21.

Scott, J. C. 1998. *Seeing like a State: How Certain Schemes to Improve the Human Condition Have Failed*. New Haven and London: Yale University Press.

Talukder, B. and D. Shamsuddin 2012. Environmental Impacts of Flood Control Drainage and Irrigation (FCDI) Projects in a Non-Irrigated Area of Bangladesh: A Case Study. *The Journal of Transdisciplinary Environmental Studies* 11(2): 16–36.

Thompson, P. M. and P. Sultana 1996. Distributional and Social Impacts of Flood Control in Bangladesh. *The Geographical Journal* 162(1): 1–13.

Toye, J. 1987. *Dilemmas of Development: Reflections on the Counterrevolution in Development Theory and Policy*. Oxford: Basil Blackwell.

——— 2013. Spending in Bangladesh: The Most Bucks for the Biggest Bang. http://www.economist.com/blogs/banyan/2013/02/spending-bangladesh, accessed January 23, 2014.

United Nations Development Programme (UNDP) 2004. *A Global Report: Reducing Disaster Risk a Challenge for Development*. New York: UNDP.

Wallerstein, I. 2004. *World System Analysis: An Introduction*. Durham and London: Duke University Press.

Wood, G. D. 1984. Provision of Irrigation Services by the Landless – An Approach to Agrarian Reform in Bangladesh. *Agricultural Administration* 17: 55–80.

World Bank 1998. *Bangladesh – Gorai River Restoration Project (GRRP)*. Washington, DC: World Bank. http://documents.worldbank.org/curated/en/1998/10/439608/bangladesh-gorai-river-restoration-project-grrp, accessed May 23, 2013.

4 Agricultural transformation at Chapra

As a result of government policies and interventions the agricultural system at Chapra is being transformed from an ecocentric to a technocentric system. I have indicated the nature of some of these changes in chapters two and three but, in this chapter, I describe them in greater detail. They include a wholesale shift to new crops, new cropping schedules, more reliance on export markets, increasing use of chemical fertilizers and greater reliance on capital-intensive technologies such as tube-wells. They also involve increasing levels of commodification of water, land and agricultural products including seeds, fertilizers and crops. Due to the differential socioeconomic capabilities of farming households at Chapra, the more wealthy households are generally able to benefit from these changes whereas the marginalized households face further marginalization.

The pattern of agricultural transformation I describe in this chapter is similar to that described for many other settings during the period of the green revolution (Blair 1978; Hossain 1988) and is continuous with colonial policies intended to transform indigenous agricultural systems in the developing world into 'modernized' systems. As many scholars have noted, this transformation leads to the loss of local ecological knowledge and its replacement by scientific knowledge and the commodification of 'nature' (Baviskar 2007; Cleveland and Murray 1997; Escobar 1996; Kottak 1999; Mehta 2001). James Scott (1998:241) points out that, according to colonial officials, "the practices of African cultivators and pastoralists were backward, unscientific, inefficient and ecologically irresponsible. Only close supervision, training and, if need be, coercion by specialists in scientific agriculture could bring them and their practices in line with a modern Tanzania." Science, in this context, is controlled by corporate actors, not by scientists (Latour 2004). States become political entrepreneurs promoting the production of scientific knowledge rather than indigenous knowledge consistent with neoliberal development agendas (Brandes 2005; Faber and McCarthy 2003:39; Guha 2000:4; Hanson 2007:599–600; Mascarenhas 2007:566; Robeyns 2005:94–95). The general outline of this process of agricultural transformation has been documented for Bangladesh by other scholars (Islam and Atkins 2007; Paul 1984; Sillitoe 1998:204; Zaman

1993). In this chapter I provide documentation of this process for Chapra specifically, on the basis of information gathered directly from local farming households.

Cropping patterns

As described in chapter two, there are three major cropping patterns known as kharif-1, kharif-2 and robi that correspond to three distinct agricultural seasons which I describe here as the spring season or the summer, rainy season and the winter season. The kharif-1 crops are grown from March to May immediately after the winter season and before the rainy season. The kharif-2 crops are grown during the rainy season from June to October. The robi crops are grown during the winter season from November to February. As noted previously, the rainy season is called borsha and the dry months of both winter and spring season are called khora, terms that are best understood as agro-ecological in meaning, since they designate cropping patterns as well as seasonal variations. My focus group discussants and case study households pointed out that this seasonal cycle has changed, however, since the construction of the Farakka diversion.

Formerly, the kharif-1 season would begin with the arrival of the first spring rains that would bring the dry, hot weather of the late winter season to an end. Farmers would plant and harvest crops during this time that were suitable for this particular season. Some of the traditional kharif-1 crops are muskmelon, watermelon, gourd, chili, spinach, okra and cucumber. Farmers have followed this seasonal cropping pattern for generations. Kharif-1 crops can neither survive in extreme drought nor flood conditions. However, the kharif-1 cropping pattern no longer occurs in the same way. Kamal and Ibrahim, focus group respondents, informed me that early floods inundated the kharif-1 seasonal crops – like oil seed, groundnut, chili, vegetable, okra and bitter gourd – at Chapra in 2011. This occurred through a combination of heavier than normal rainfall and the fact that India stopped diverting water at the Farakka point, thus causing an abnormal and unexpected surge of water downstream in Bangladesh. In other years, rainfall amounts have been exceptionally low, causing drought conditions that can in turn create virus and insect problems for many kharif-1 crops. In these cases, India diverts a larger than normal percentage of Ganges water to Kolkata, creating an extreme drought situation in Bangladesh, which otherwise might have been manageable. Since the increasing incidence of both flood and drought are causing a decrease in the production of traditional local kharif-1 crops, many farmers are now growing the government supplied HYV crops during the kharif-1 season.

The kharif-2 season begins immediately after kharif-1. During the kharif-1 crop harvesting period, farmers begin to prepare croplands and kharif-2 seedlings before the monsoon rain of the borsha season begins. Formerly most cropland was planted in two main kharif-2 crops, aman paddy and jute, which traditionally were the foundation of the national economy in Bangladesh. Aman paddy rice was a staple food that people in Bangladesh would eat three times a day. Jute was called the golden fiber because it was the main source of foreign currency. However, the kharif-2 season is facing seasonal irregularities today due to reductions of the Ganges River flow. The disappearance of local water bodies and the associated lowering of water tables make it difficult to produce both aman paddy and jute. Kamal also informed me that bull frogs locally called *kola bang* no longer herald the rainy seasonal resumption, since they are disappearing from local water bodies due to ecosystem failures. Finding no other alternatives, local farmers plant the government-promoted HYV rice paddy for the kharif-2 season just as they do for kharif-1.

Tanvir, Tofajjal, Joardar and Billal, in-depth case respondents, argue that the rainy seasonal failures in turn generate major challenges for the robi crops during the khora season. The reduced water levels in the Ganges and Gorai Rivers lower local water tables, thus reduce the level of water in local water bodies at Chapra, like Lahineepara Beel. As a result, they often do not have the minimum water necessary for robi crop production. Under normal conditions, the level of water in Lahineepara Beel is such that it raises the water table and plants are able to capture some of this water without irrigation. But respondents Fazlu, Abonti and Bimal state that the increasing dryness of the land creates higher temperatures, which cause a higher incidence of medical problems, such as heat stroke, as well as crop failures.

Farming families face major challenges when attempting to follow traditional seasonal cropping patterns. As the khora season length is increasing, due to different factors like the Farakka diversion, local wetlands face a longer shallow or dry time, which causes challenges for producing the kharif-1 and kharif-2 crops. To overcome these challenges, farmers produce one new rice paddy crop locally called boro at the end of kharif-1; boro production has been dominating cropping patterns since the 1990s. For example, boro was planted on 76,890 hectares out of a total of 236 thousand hectares in Kushtia in 2004–05 (Figure 4.1). When boro is planted in this way, there is not sufficient

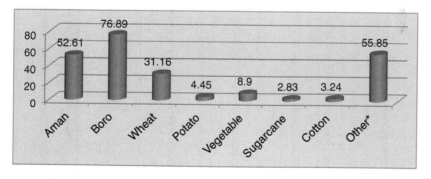

Figure 4.1 Crop types (in thousands of hectares) in 2004–05 in Kushtia District

Source: Bangladesh Bureau of Statistics (2005)

time left for producing kharif-2 crops like aman. Furthermore, the rainy season failures, since they lower the water table, cause winter season failures that reduce production of robi crops like wheat, pulse, lentil, onion, chilly, vegetable, potato, cauliflower, radish, cabbage, tomato, mango, eggplant and okra. Reductions of wheat crop production also occur and are especially significant because this crop is the second in importance to the paddy crop for household food self-sufficiency. In 2004–05, farming households in Kushtia District planted 31,160 hectares of wheat (Figure 4.1). The continuous ecosystem failures caused by reduced flows past the Farakka diversion are hidden by the fact that overall production figures remain high. By paying attention only to overall production figures, government analysts fail to recognize the ways in which their interventions are actually changing seasonal patterns, causing environmental degradation and increasing marginalization of smallholders. My in-depth case respondents termed these production challenges as *songkot*, or "crisis," originating in the Ganges flow reduction.

Focus group respondents explained that they are unable to maintain traditional land use patterns because of the Farakka diversion. Jalil argued that the diversion creates major challenges for the traditional agro-ecological system in all elevations within the Ganges River floodplain in Bangladesh. Currently, the higher elevation croplands fail to receive river water, except during an extreme flood event, and, therefore, lose ecosystem-based fertilizers like earthworms, water hyacinth and siltation, resulting in lost soil fertility. Mid-level croplands often encounter either drought or water stagnation because of river flow irregularities. Lower croplands encounter the worst problems due to flooding, drought, water stagnation and embankment failures. These ecological problems create agricultural production challenges for the robi, kharif-1 and kharif-2 cropping periods.

Tarun pointed out that the Ganges flow reduction generates ecological knowledge failures in agricultural production. Chapra communities fail to practice the higher, intermediate and lower croplands in coordination with cropping and seasonal patterns. Tarun and Kabir's deeper understanding about the kharif-2, robi and kharif-1 cropping patterns in coordination with the rainy, winter and spring seasons respectively is no longer applicable, due to the river flow reduction and disappearances of local water bodies. Alagmir and Gedi argued that cropland fertility practices and knowledge about siltation, algae and water hyacinths are no longer applicable in agricultural production and, as a result, croplands are losing productive capabilities. Intercropping practices are no longer applicable because of the changing and unpredictable characteristics of the seasons.

Kafil argued that the Ganges flow reduction causes traditional agricultural production failure in *vita* systems. This system involves building a raised bed to grow crops like eggplant, chili, beans or okra. Because of the slightly higher elevation of vita crops, they remain safe from water inundation for a longer time than crops grown in the normal manner. Kafil and Rahim also use this vita system to produce seedlings so they can begin crop production immediately after rainy season water recedes from croplands. However, use of these vita systems is diminishing day by day because of the Ganges River ecosystem failures.

The river bank communities at Chapra are also no longer able to practice the traditional *maacha* cropping systems. Maacha systems require building a platform with natural resources like bamboo, sticks and rope on a relatively higher level of a cropland to secure crop production during the rainy season. Joardar and Billal practice this maacha system for producing crops like bitter gourds, thus somewhat overcoming flooding or water stagnation problems. Moreover, they use the rooftops of residential houses to produce vegetables such as pumpkins. They plant seeds in a household gardening plot and connect the plants to rooftops using bamboo sticks. However, these maacha and rooftop systems are diminishing in use due to the current irregular climatic characteristics. Changes in the seasonal characteristics of the region thus lead to the failure of many traditional cropping practices, such as vita and maacha and other practices that traditionally allowed for the effective utilization of croplands at all elevations. In the short term, this results in lower crop productivity and in the long term it will also result in the permanent loss of the local ecological knowledge associated with traditional agro-ecological practices. Alamgir and Bakar argued this loss as: *"sorkar amader nijaswa shikor tule phelchen borolokder sharthe"* (the government is destroying our roots for elite interests).

Crops and seeds

Indigenous crop varieties

My focus group respondents, notably Kusum, argued that the reduced flows of water in the Ganges River system are responsible for the reduced productivity and viability of local crop varieties. Joardar and Abul described how they would traditionally select the best portions of their croplands and scrutinize their elevation, temperature and soil quality so that they could produce the best possible quality of seed. Not all seeds for the next year would come from specialized plantings, however; Abul and Faruk also reported preserving some seeds from regular crops when these crops pass their seed quality standards. Based on these traditional activities, they would retain seeds like paddy, onion, ginger and turmeric for the kharif-1, kharif-2 and robi seasons.

Suma, Joardar's wife, produces local seeds like beans and watermelon. She has detailed traditional ecological knowledge about the specific temperature necessary to preserve these seeds. Suma maintains this temperature with traditional systems. Joardar and Suma preserve seeds like onion in an earthenware jar locally called *kolosh* to ensure an ecosystem friendly setting that allows no extreme hot or cold. Suma inspects the temperature and seed condition of this jar every 15 days. However, the Ganges flow reduction creates major challenges for traditional seed production when it causes greater extremes of temperature or humidity. The failure of seed production in turn creates major challenges for traditional agricultural production. As Joardar, Rahim, Harun and Laily explained, they need to estimate their total seed demands one year ahead for each cropping season and preserve kharif-1, kharif-2 and robi seeds for a year. This is difficult enough under normal circumstances but much more difficult today when technological interventions in the hydrological system create major seasonal unpredictability and turn manageable episodes of flooding and drought into more extreme, unmanageable events.

During my focus group discussions, Tofajjal noted that traditional seed production practices involve the sharing and exchange of seeds and thereby the strengthening of community bonds. Keramot and Harun pointed out that the concept of selling seeds as a commodity was never part of their traditional agricultural practices; and sometimes, they still exchange seeds like paddy or eggplant, based on good wishes and community sentiments. Some other seeds, like pumpkin, are exchanged freely among neighboring women. Salma, Parvin and Suma, wives of Tofajjal, Billal and Joardar respectively, share these seeds without any cost so they can produce seasonal vegetables. They also share the produced vegetables freely to generate good wishes and develop community bonds. However, the volume of sharing has been reduced as a result of recent natural resource and farming practice changes. Reductions in vegetable production also reduce the food previously available for domestic animals, a topic discussed in more detail in the following section. My survey data indicate that 87 percent of the local population is facing major challenges to their practice of these traditional cropping patterns. The displacement of these cropping patterns have forced many farmers to accept the HYV crop production, finding no other alternatives.

Introduced crops

The Ganges Basin ecosystem failure is a main reason for accepting seed commodification, which creates the differential outcomes for the rich and marginalized farmers at Chapra. Kafil explained to me how local seed practice failures create external seed dependencies for agricultural production. Kafil with Gedi and Bimal, two other focus group respondents, informed me: *"amader prakrtir sathe taal milie shosway chaas bondo hoyea jacche bideshi shoswayr karone"* (our bond with nature for crop production is decreasing due to foreign crops). The problem begins with the central government's policy to promote HYV seeds and ignore the marginalized farmers' voices about local seed species. This top-down seed policy promotes seed commodification, since it is now impossible for farmers to produce sufficient local seeds. Better-off farmers, like Tanvir and Tofajjal, are able to afford the new seeds and have the capital and resources necessary to grow them successfully; whereas, the marginalized farmers, like Joardar and Kafil, do not have economic capability and thus face major socio-economic challenges with lowered agricultural production.

The central government provides subsidies for the HYV seeds. In the case of rice, the government introduced HYV aus, HYV t-aman and HYV boro

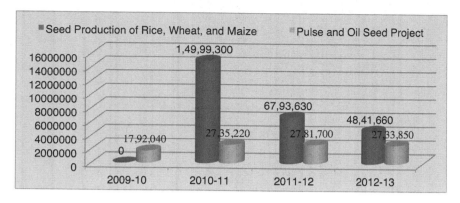

Figure 4.2 Two seed production program budgets in Bangladesh
Source: Ministry of Finance (2012)

after increasing their seed development budget 510 percent from US$8,154,140 in 1972 to US$497,403,000 in 2006[1] (Ministry of Agriculture 2007). Figure 5.2 displays the government budget[2] to improve just five crop types (rice, wheat, maize, pulse and oil seeds) from 2010 to 2013 (Ministry of Finance 2012). These budget allocations are dominated by different factors like government policy priorities. This greater budget allocation for HYV seed development led to increases in the planting of new seed varieties of aus, aman and boro by 1085 percent, 541 percent and 2983 percent respectively between 1990 and 2003 (Department of Agricultural Extension 2007). These HYV seeds dominate local agricultural production. On the other hand, production of indigenous local seeds, like aus, t-aman and b-aman fail to receive government support, and therefore, local people fail to produce them due to ecosystem failures.

The government seed policy promotes private sector involvement in seed development and marketing and this worsens the differential outcomes between the rich and marginalized. Currently, local seed businesses are a new profit-making opportunity for the rich farmers. However, the marginalized farmers are excluded from this business; consequently, they turn into seed consumers of local seed business farms, and national and international seed industries. Since 1998, the central government has permitted private companies to produce HYV seeds (Table 4.1). Many of these companies use their own marketing systems to sell their own new varieties of seeds. For example, BRAC, a major NGO in Bangladesh and one of the largest in the world, encourages their clients to buy its own HYV seeds. Additionally, the central government permits local and international companies to import the HYV seeds from abroad and sell them in local markets.

Table 4.1 Privatization of seed production in Bangladesh

Year	Organization	Hybrid Rice Seed names
1998	BINA (BADC)	BINA 4, BINA 5, BINA 6
1998	BRRI (BADC)	BRRI dhan34, BRRI dhan35, BRRI dhan38,
1998	GDC	Amar Siri 1
1998	McDonald Bangladesh (Private) Limited	Loknath-503
1998	ACI Limited	AILOK-6201, Allok-93024,
1998	Mollika Seed Company	CNSGC6(Sonar Bangla 2), HTM-4(Sonar Bangla-6)
2003	Supreme Seed Company	HS273
2003	BRAC	GP4
2006	National Seed Company Limited	Taj 1(GRA2), Taj 2(GRA3)
2006	North South Seed Limited	HTM-606, HTM-707
2006	Sea Trade Fertilizer Limited	LP 108
2006	East West Seed Bangladesh Limited	HTM 202, HTM 203
2006	Sungenta Bangladesh Limited	LU YOU-3(Surma 2), LU YOU 2(Surma 1)
2006	National Seed Company Limited	Taj 1 (GRA 2), Taj 2 (GRA-3)

Source: Ministry of Agriculture (2007)

Joardar pointed out how the seed development programs have increased the cost of local seeds. In 2005–06, 1 acre of cropland required spending an average of US$9 a year for HYV t-aman, HYV boro and HYV aus rice seed (Figure 4.3). The cost of the equivalent local seeds was higher than the HYV seeds because of government seed policy. The central government provides major subsidies for the HYV seeds, while no subsidies are given to the local seeds that are now produced in lesser amounts and are harder to come by. Therefore, the marginalized farmers like Joardar and Billal cannot afford to buy local seeds when they are unable to produce their own and also find it difficult to afford the subsidized seeds. However, the rich farmers like Tanvir and Tofajjal benefit both as commercial agricultural producers and through involvement in the seed business.

My in-depth case respondents, Tanvir, Tofajjal, Joardar and Billal, are thus all increasingly depending on the HYV crops. My household survey data indicates that 60 percent of the farmers were using these HYV seeds in 2010. In 1969–1970, the HYV aus, aman and boro crop production was 1,397 metric

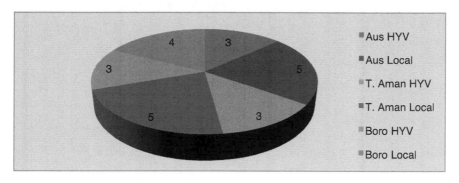

Figure 4.3 Subsidized seed costs (US$) for 1 acre of cropland in 2005–06
Source: Department of Agricultural Extension n.d.

Table 4.2 Comparison of HYV and indigenous crop production (metric tonnes) between 1967–70 and 1999–2000 in Kushtia District

Year	HYV Aus	Broadcast Aman	HYV Aman	HYV Boro
1969–70	1,397	51,885	36	1,910
1999–00	40,570	5,490	253,080	242,970

Source: Adapted from Chowdhury and Zulfikar (2001:88, 90, 109, 111, 121, 123, 151 and 152)

tonnes, 36 metric tonnes and 1,910 metric tonnes respectively in Kushtia but this increased to 40,570 metric tonnes, 253,080 metric tonnes and 242,970 metric tonnes respectively in 1999–2000. On the other hand, the indigenous rainy season dependent broadcast aman production fell from 51,885 metric tonnes in 1969–70 to 5,490 metric tonnes in 1999–2000 (Table 4.2). This local crop production decrease and the HYV crop production increase promotes profit for the rich farmers but produces major agricultural and survival challenges for the marginalized farmers.

In addition to this market economic effect, ecological vulnerabilities like flooding and drought create the differential outcomes for the rich and marginalized farmers when growing HYV crops. These crops require strictly controlled watering schedules and cannot survive with water extremes of flooding and drought. The Ganges Basin farmers encounter sudden, unpredictable flooding, earlier or later flood periods, and more periods of drought, all of which are not suitable for these HYV crops. Joardar and Kafil cannot cope with these environmental vulnerabilities because of their limited land ownership and economic resources. However, the rich farmers are successful in overcoming these vulnerabilities based on their larger and varied cropland holdings and capital resources that allow them to purchase the water which is now available as a managed commodity rather than a free resource.

Water commodification

The reduced volume of water in the Ganges and Gorai Rivers is responsible for many of the hardships now being faced by poor farming households but the commodification of water, made possible by the Ganges-Kobodak Project and the proliferation of tube-wells, is also the cause of new forms of hardship. The central government has introduced these technologies following the advice of the World Bank, the International Monetary Fund and other international agencies, in the hope of addressing the issues of hunger and environmental sustainability (Water Resources Planning Organization 2009:9). The benefits of these technological innovations are almost entirely captured by small elite, well-off farmers. Sabbir and Tarun identified this problem with the comment: "*paani bebosa amader sob sommossar mul*" (Water commodification causes other problems). They elaborated further: "*Paani prakritir daan athocho donira eta die babosa kore*" (Water is a gift of nature for all but the rich makes profit by controlling it.)

As outlined in chapter three, those who want to obtain water from the government-operated GK Project must purchase it and this creates unequal access for rich and poor. Much more inequality has been created, however, through the privatization of water supplies by construction of tube-wells. This form of privatization is facilitated by article 4.7 of the 1999 National Water Policy, which specifically addresses irrigation privatization for agricultural production. This is accomplished mainly through the construction of shallow and deep tube-wells by farmers themselves, on their own properties, so they can withdraw water from local aquifers. The number of shallow tube-wells increased by 5,307 percent between 1979 and 2005; and deep tube-wells increased by 296 percent during that time period (Bangladesh Bureau of Statistics 2005). Based on these irrigation projects, rich farmers are able to gain greater control over local water management. Tanvir has two shallow tube-wells and one deep tube-well in Kumarkhali sub-district.

The government provides subsidies and tax exemptions for the purchase and installation of tube-wells on the basis of various development programs, including structural adjustment plans, the National Poverty Reduction Strategy Paper and the Millennium Development Goals as adopted by Bangladesh. Local rich farmers, like Tanvir, own the wells as private property and operate them without any effective restrictions or regulation. According to Tanvir, he was initially required to obtain a license to install and use the wells but he is not required to renew them periodically. He can withdraw as much groundwater as needed for irrigating his own lands and he can charge as high a water fee as he can get when selling water to others. Tube-well owners can also sell its water to local clients without restriction.

Joardar and Billal do not have the economic capacity to purchase this technology and often have to purchase their water from the rich farmers. Joardar and Suman stated that Tanvir controls the price of water supplied in this way and this control over water has become an important component of how local power structures are maintained in water resource management.

The domination of this power structure along with cost of irrigation water constrain agricultural production among poor households. According to the Department of Agricultural Extension, the average annual irrigation cost per acre for HYV and boro rice crops was US$87 in 2005–06 (Figure 4.4). This includes the cost of water delivered from all sources, including GK water and private tube-wells. This total cost includes irrigation water for aus HYV, t-aman and boro HYV but does not include irrigation costs for robi crops like wheat. Since most of marginalized farmers, who make up 93 percent of the population in Kushtia, have been forced by circumstances to grow HYV crops, irrigation water expenses have become a significant new economic burden for this group. Joardar often fails to receive water supplies he is entitled to from the GK project and becomes vulnerable to exploitation by Tanvir since he controls the local GK project water management association and extends his control of the local water supply through his ownership of shallow and deep tube-wells. These tube-well projects work independently of the GK Project system. They have their own canal infrastructure for water transportation and drainage, separate from the GK canal infrastructure. Tube-well owners like Tanvir control water supplies, service charges, canal construction and repair works for their own projects. In this context, community egalitarian water practices are replaced by monopolized market economic systems. The basin communities at Chapra are failing to resist these systems due to the Ganges flow reduction.

The decreasing of river water causes increasing dependency on rich people at Chapra. Kamal and Ibrahim reported spending a significant amount for

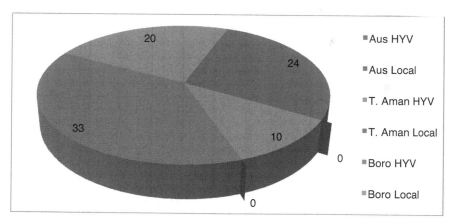

Figure 4.4 Irrigation costs (US$) for 1 acre of cropland in 2005–06 in Kushtia

Source: Department of Agricultural Extension n.d.

domestic and irrigation water. Earlier, they used the Gorai River and local water bodies for domestic activities and dug ring-wells or *kuya* for drinking water, domestic supplies that were readily available without any expenditure. However, due to lower water tables, these sources are no longer available. Tanvir has two tube-wells, one inside and another outside of his residence. On the other hand, Billal and Joardar do not have any tube-wells and need to seek permission from Tanvir for tube-well water for drinking and domestic activities.

New technologies increase available alternatives to the rich people but decrease the previous opportunities to the marginalized people at Chapra. In addition to tube-well water, Tanvir buys bottled water when his son, Shaon, and grand-children, who now live in Australia, come for a visit. There are numerous private companies in Bangladesh producing bottled drinking water at local markets. Tanvir's grandchildren always drink this bottled water when soft-drinks like *Coca-Cola* are not available at Chapra. Parvin, Billal's wife, sees these practices during her work as a domestic helper in Tanvir's house and feels frustrated for her own children and grandchildren. Billal fails to get tube-well water and sometimes drinks pond or wetland water that has been contaminated by chemical fertilizers, which can cause diarrhea, dysentery and skin diseases. Clearly, this differential water access is a major outcome of water commodification and an important cause of reinforcing social inequality at Chapra.

Fisheries commodification

According to Bimal, *"pani jokhon theke amader kinte hocche tokhonoy anno sob kichur bechakena suru"* (the beginning of water pricing introduce commodification of other goods). Water commodification has led in turn to fisheries commodification. Wild fisheries are failing due to reduced river flows and lowered water tables and in response to this the local elite develop commercial fisheries projects, thus creating a new business out of this ecological crisis. As the Gorai River and its dependent local water bodies disappear, many fish species are becoming extinct. The central government's dredging operations create further challenges for fish species like eels (*kuichya*) and snakehead fish (*gajar*). Furthermore, chemical fertilizers and pesticide contamination in local water bodies destroy local fish species like the minnow (*puti*) and also crocodiles (*kakla*) at Chapra. Khan (2013) informed us that more than 28 indigenous fish species have became extinct due to interventions such as the Farakka diversion and ecological degradation. Currently, 10 percent of 260 local fish species are reported to be encountering survival challenges (Khan 2013). Ninety-three percent of my household survey respondents reported significant reductions in their ability to obtain fish for family consumption or for sale in local markets.

Fish commodification reduces its consumption to the marginalized people at Chapra. Sabina informed me that she can only afford to buy lower priced fish like tilapias and Thai pangasius. She and her family encounter major challenges even to buy lower priced fish since 1 kilogram of tilapia costs US$1.50 in the market at Chapra. Sabina informed me that a household with five members needs 157 kilograms of fish for a year, which requires spending US$204. This amount is beyond their economic capabilities and so they reduce their fish consumption. Islam et al. (2004a) report that a landless household in this region consumed on average 157 kilograms fish in 1990 but reduced to as low as 58 kilograms in 1999.

Local fish species extinctions have been followed by the introduction of new fish species suitable for commercial production. For these purposes, the government allocated US$179 million in 2013 (Ministry of Finance 2013:74). The Bangladesh Fishery Research Institute has developed new fish varieties, like Thai climbing perch (*gugi*) and catfish (*nona tengra*) (Ministry of Finance 2013:74). The central government has also supported the introduction of foreign fish species like carp, tilapia and Thai pangasius which in turn serve to create new fisheries business opportunities for the rich farmers at Chapra.

Joardar and Billal argued that rich people like Tanvir control local fisheries projects and that these projects establish their justification that local fish populations will not recover and that fewer fish will be available for consumption by poor families or as a source of traditional fishing employment. Local political leaders of the ruling government, in coordination with rich farmers, often use local water bodies to develop their fisheries projects. Political leaders even blocked the Ganges River for 2 kilometers downstream from Kushtia to develop a major commercial fisheries project (*Prothom Alo* 2013b). At Chapra, Tanvir has blocked the Kaligangya River to establish one of his fish projects. In addition, he obtained leases for two other fisheries projects from the GK project by exploiting the local power structure. As I discuss in more detail below, commercial fish farms also provide a means of transforming local employment practices from ones that were relatively autonomous, such as catching wild fish on one's own schedule, to practices that involve a greater dependency on elite families of poor households and thus more opportunity for their exploitation.

Land and equipment costs

Households without sufficient land to support the family are forced to rent land or work as day laborers. According to my survey data, 49 percent of the farming households at Chapra depend on sharecropping for a portion of their livelihoods and are forced to pay high rents for this land to the wealthy land owners (who are the only people who have excess to land). Land commodification was less common in the past when land scarcity was less prevalent and rich people often shared croplands with poorer households based on principles of reciprocity, kindness and social justice. The commercialization of all aspects of the agricultural economy makes such arrangements highly unlikely today, however. In 2005–06, land rental charges for boro HYV, HYV t-aman and HYV aus crop production, for one cropping period, were US$34, US$29 and US$25 respectively for 1 acre of land (Figure 4.5).

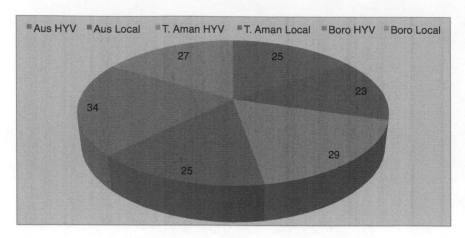

Figure 4.5 Land rental costs (US$) for 1 acre of cropland in 2005–06
Source: Department of Agricultural Extension n.d.

Sometimes, as an alternative to sharecropping, land poor households obtain *kot,* or loans from an employer or other wealthy individual, a practice that creates further dependency. However, this practice is currently following the opposition direction and the rich farmers are exploiting it for further benefits. In 2001, Joardar received US$1,000 from Tanvir in exchange for 0.33 acres of cropland. Trust and tradition between them were the two major criteria for this *kot* and Joardar made a verbal commitment to return the money within one year. However, Joardar failed to return the money due to a combination of flooding and then drought that occurred during that year. Later on, he borrowed an additional US$500 but failed to recover from his economic losses. He eventually had to sell this land to Tanvir in 2007 for US$3,000 although its market value was US$6,000.

All households in Chapra today also face additional costs to purchase farm equipment that can no longer be made locally because of a lack of materials like wood. There is insufficient wood, for instance, for making boats, bullock carts, winnow fans and husking pedals. To overcome these shortages, the national government subsidizes tractors, transportation vans, boats and paddy husking machines that are imported with assistance from the World Bank (International Monetary Fund 2013). The government permits private companies to import these technologies and to sell them directly to local farmers. The central and local governments are officially responsible for ensuring accountability throughout this process. The shortage of locally manufactured farm equipment means that many poor farming households have to rent this equipment. Joardar, for instance, rents a tractor and plow from Tanvir once Tanvir has finished using it to prepare his own fields (Figure 4.6). Many poorer households cannot afford

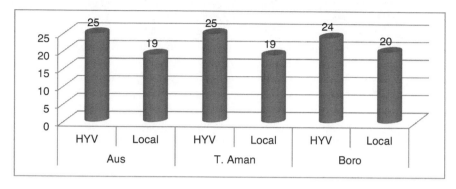

Figure 4.6 Land preparation costs (US$) with tractor and machine for 1 acre of cropland in 2005–06

Source: Department of Agricultural Extension n.d.

to rent tractors and have to do the work manually. Billal and his family, for instance, prepare their field with hoes; his son, Jamal, and wife, Parvin, will then shoulder the yoke and try their best to pull the plow that would normally be pulled by a bullock. Ninety-three percent of my survey participants report practicing this type of cultivation system. Tanvir, on the other hand, is able to earn US$1,000 by renting his tractors two seasons in a year.

Fertilizers and pesticides

As described in chapter two, Chapra farming households would traditionally benefit from several forms of freely available "natural" fertilizers such as siltation, algae, earth-worms, water hyacinths and animal manure. However, most of these natural fertilizers are available today only in much reduced quantities or are not available at all, due to changes in the ecosystem characteristics of the region. As a result, farmers need to purchase chemical fertilizers or suffer significant losses in crop productivity.

Natural fertilizers

Focus group participants Hanif and Naser described the ways in which they would obtain natural fertilizers from the Gorai River and Chapra *beel*. During a regular borsha season, the Ganges River flow carries silt that would cover all croplands connected to the Ganges River and its tributaries by local wetlands, canals and oxbow lakes. This siltation is a major source of soil nutrients and no other fertilization is needed. However, since the Gorai River is itself encountering survival challenges due to the Farakka diversion, many local wetlands and ponds no longer exist at Chapra. As pointed out in chapter three, the Gorai

River is filled with sedimentation and charland that prevent the replenishment of local bodies of water during the rainy season.

Kamal demonstrated a particularly detailed understanding of ecological resources like algae, which are declining due to the river flow reduction. He informed me that local wetlands, lakes, canals and croplands themselves will produce algae as long as Gorai River water is available in sufficient quantity during the rainy season. This algae grows naturally in croplands and water bodies during this season, which keep croplands fertile and increases crop production. All farmers like Joardar and Morjina receive these ecological services equally, irrespective of their socioeconomic hierarchies. However, this algae dependent cropland fertility has diminished significantly due to the reduction of flow in the Ganges and Gorai Rivers.

Water hyacinth is another source of cropland fertility that is reducing day-by-day because of the Ganges River ecosystem failures. This hyacinth used to be readily available in local water bodies and was easily movable from one place to another during the rainy season. Joardar would pile up water hyacinths in croplands at the end of the rainy season and make them into green fertilizer through traditional methods. The hyacinths were allowed to compost for about 2 months before they were applied to the fields. Sometimes, local farmers would place water hyacinth compost on their croplands to retain soil humidity for a longer time during the winter season for the robi crop production. They would also use them as cooking fuel. In the past, the plant was traditionally abundant as well as free; however, this water hyacinth is no longer available in most of its previous habitat due to the reduced flow of water throughout the system. Much of local cropland at Chapra is also disconnected from local streams, ponds and wetlands due to interference by canals and other water irrigation infrastructure failures.

Earthworms are another traditional source of fertilizer but are much diminished in numbers due to the river flow reduction. One particular species of earthworm, locally called *kecho*, used to be abundantly present in croplands as a natural consequence of the local ecosystem. My focus group respondent Barek reported seeing a greater number of kecho every year at the beginning of the rainy season. This kecho burrows into topsoil, eating organic matter and excreting castings which enhance soil fertility. This fertility source is freely available for every farmer at Chapra without discrimination. River fish like snakehead also eat kecho and local fish production also therefore depends on its availability. However, according to my informants, the Ganges River flow reduction has significantly decreased the number of kecho in croplands as this requires a specific ecosystem connected with river water and croplands. Moreover, contemporary excessive usage of chemical fertilizers and pesticides further reduces the kecho population in croplands.

Harun is now unable to obtain animal manure fertilizers due to a reduction in domestic animal populations. Earlier, the Ganges flow maintained natural resources like grasslands and forests that were helpful to raise these animals. Harun and Kabir described how they used to gather dung from these areas and deposit it in pits on their homesteads for 11 months until it had broken down sufficiently to be used as manure that they would deposit on their croplands at the beginning of the winter season. This would ensure cropland fertility throughout the robi, kharif-1and kharif-2 cropping seasons. No expenditure was involved in making this manure and everybody could make it equally without any discrimination. However, these manure fertilizers, in addition to water hyacinth, siltation and earthworms are diminishing day-by-day due to the Ganges River flow reduction.

Chemical fertilizers

To overcome the loss of ecological fertilizers, the central government supported the creation of more than 10 major fertilizer industries and established an import program for fertilizers from abroad. As in the case of the HYV seed industry, well-off farmers can afford to purchase these fertilizers and are also able to create new businesses based on their sale. However, the marginalized farmers face extra economic burdens and survival challenges because of this commodification. In 2006, a total of 3,551,000 metric tonnes of fertilizers such as Triple Super Phosphate (TSP), Single Super Phosphate (SSP), Diammonium Phosphate (DAP), Muriate of Phosphate (MP), Ammonium Sulfate Phosphate (ASP) and Nitrogen-Phosphorous-Potassium-Sulfur (NPKS) were purchased for crop production in Bangladesh. This represents an increase of 306 percent between 1980 and 2006. In Kushtia, 69,155 metric tonnes of fertilizers of urea, TSP, DAP and MP were used in 2011 on 348,000 hectares of cropland (Table 4.3). The rich farmers are the major beneficiaries from these fertilizer usages and sales. Kofil, Tanvir's son, has a fertilizer business and Tanvir promotes commercial crop production based on these chemical fertilizers. On the other hand, the marginalized farmers like Joardar encounter major economic challenges for affording these fertilizers in agricultural production and they do not have scope for establishing fertilizer business.

Chemical fertilizer use represents a highly significant extension of agricultural commodification. Currently, 1 acre of cropland requires an average of US$90

Table 4.3 Fertilizer usage in Kushtia

Kushtia	2008–09 (in metric tonnes)	2010–11 (in metric tonnes)	Increase (percentage)
Urea	46,608	50,898	9
TSP	1,561	6,971	347
DAP	524	7,196	1,273
MOP/MP	923	4,090	343
Total		69,155	

Source: Ministry of Agriculture (2012)

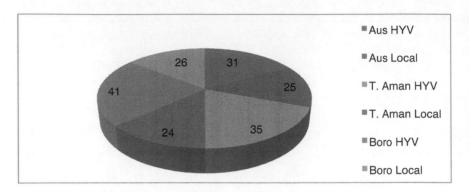

Figure 4.7 Fertilizer costs (US$) for 1 acre of cropland in 2005–06
Source: Department of Agricultural Extension n.d.

a cropping year for fertilizer expenditures to produce HYV aman, HYV t-aman and HYV boro or irrigated wheat crops (Figure 4.7). In this context, 1 acre of cropland requires US$41, US$31 and US$35 a year for fertilizer expenditures in HYV boro, HYV aus and HYV t-aman crop production respectively. Rich farmers like Tanvir and Tofajjal can afford these fertilizers for commercial crop production. They are the major beneficiaries of the central government's fertilizer subsidies, and they have the capital and government connections necessary to start a fertilizer business. On the other hand, the marginalized farmers like Joardar cannot afford these fertilizer expenditures because of their economic crises.

Kamal argued that chemical fertilizer usage is a main cause of loss of soil fertility. He also encounters increasing crop production failures because of new insects, fungi and weeds that appear as soon as he begins to use chemical fertilizers. To overcome these production failures, the central government introduced insecticides, fungicides and herbicides, thus further increasing the cost of crop production. Finding no other alternatives, Kamal uses these pesticides to overcome insects and fungi problems. In Bangladesh, farmers used 17,393 metric tonnes of insecticide, fungicide, weedicide and rodenticide in 2012 for these purposes. Pesticide use increased 262 percent between 1989 and 2002 in Bangladesh (Bangladesh Bureau of Statistics 2005). Currently, farmers need to spend more than US$22 per cropping year for pesticides on 1 acre of cropland (Figure 4.8). These monetary costs were not necessary when the Ganges Basin ecosystems were active at Chapra.

Despite these pesticides, farmers have not been entirely successful in overcoming insect and fungi problems. According to Joardar, insects develop resistance to many insecticides within two years but new insecticides are not made available to them within this time period. Once available, these insecticides are often more expensive and he cannot afford these increasing costs given the crop

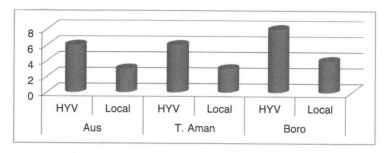

Figure 4.8 Pesticide costs (US$) for 1 acre of cropland in 2005–06
Source: Department of Agricultural Extension n.d.

production losses he experiences. On the other hand, rich farmers like Tanvir, face fewer insect and fungus problems because they are better able to access scientific knowledge necessary to these new systems of agricultural production. Pesticides have thus become one more cause for major social inequality at Chapra.

Domestic animals

Kamal explained that domestic animals are essential to traditional agricultural production but that populations are diminishing day-by-day, due to ecological resource degradations. Wetlands, grasslands and forests are needed for nurturing animals like bullocks, oxen, cows, chickens, ducks and goats. Bullocks cultivate croplands and cows provide milk for household nutrition. Traditionally, Billal and Kafil would get bullock calves from domestic cows without cost and use them to reduce their agricultural production costs. They raise these animals domestically and use wild grasslands and plants as their fodders. The bullocks were used to cultivate croplands based on their traditional agricultural practices.

Kusum explained how local ecological resources are used to nurture domestic animals. Domestic ducks feed on small fish and snails from local water bodies such as Chapra Beel. Cattle receive food from local grasslands and forest resources and vegetation like water hyacinth and banana trees during the borsha season. However, all of these free ecological resources are now in much diminished supply due to the reduced flow of water in the system (cf. Lansing 2006).

Therefore, Bakar explained how natural vegetation has been contaminated by chemical fertilizers and pesticides and how new insects and fungi destroy many forms of vegetation and diminish local biodiversity. As a consequence, Billal and Joardar reported no longer having any domestic animals like hens, ducks, goats and cattle. Many of these animals encounter health concerns and die because they eat pest affected grasses and leaves.

To recover from this domestic animal reduction, the central government in Bangladesh introduced poultry and animal farms to ensure meat supplies. For example, in 1997, the government supported the establishment of 29,649 dairy farms, 20,833 goat farms, 10,289 sheep farms, 30,760 duck farms and 60,670 poultry farms throughout the country (Rahman 2012). For this purpose, the government received assistance from the different countries like the USA, Germany, France, Australia and Japan to develop hybrid species of poultry and cattle. In every village at Chapra, rich people established more than one poultry farm. Tanvir's son-in-law has a poultry business at Chapra and the whole village depends on this farm for household meat and egg supplies. In 2013, 1 kilogram of poultry meat cost US$1.50. Based on this cost and on a daily consumption of 25 grams per person, Billal needs to spend US$123 a year to feed his nine-member family. This does not include the costs of eggs and milk. Billal and Joardar cannot afford these expenditures and, thus, go without meat for many days of their meals. On the other hand, Tofajjal and Tanvir have generous surplus supplies of meat, eggs and milk. Livestock commodification compounded by vegetation loss is thus another cause of the differential class outcomes at Chapra. The rich farmers benefit from government supported opportunities for livestock production, while the marginalized farmers face extra economic burdens as they become consumers of the rich farmers' livestock commercialization.

Kamal, Raju and Billal all reported having no domestic bullocks for cultivating croplands and thus no opportunity for getting calves from their own cows. The lack of bullocks not only limits the amount of cultivation they can do but also reduces their ability to manufacture agricultural production materials like hoes, sickles, paddy husking pedals, ploughshares and cleavers that are produced traditionally from the slaughtered animals' bones, horns and hooves. They fail to get milk from domestic cows and eggs from chickens and ducks. These domestic animal reductions and ecological resource degradations generate major livelihood challenges for the marginalized communities at Chapra.

Domestic animal reductions also damage traditional exchange relationships within the community. In 2010, Joardar borrowed a bullock from a widow and bought her groceries from local markets in exchange. In 2011, he loaned a tractor and plow to Billal for agricultural production during the summer season and received boat rides from Billal during the rainy season. These practices reduce the differences among farmers and promote community cohesion and egalitarian practices. However, these exchange practices are decreasing due to continuous increase of market dependencies.

Employment

Traditional patterns of employment at Chapra have also changed dramatically as a result of the commodification of agriculture and the inability of marginalized households to produce sufficient food through traditional methods. Among my survey respondents, 56 percent reported significant periods of unemployment in 2011 in the agricultural sector or in the other occupations such as boatmen, potters and fishermen who depend directly or indirectly on agricultural productivity. Women, the elderly and the disabled face the worst employment challenges in Chapra.

Local agricultural employment practices have the different socioeconomic dynamics. The intermediate farmers like Barek perform supervisory tasks for more wealthy farmers like Tanvir, while landless day laborers like Billal work for Tanvir in agricultural production activities such as cultivation and irrigation. The day laborer is locally called *kamla,* and this individual generally works for the rich farmer from dawn to dusk and is fed three times a day as part of an informal contract. The small farmers like Joardar and his wife, Suma, mainly work on their own land but also work as day laborer for more wealthy landowners on occasion. Joardar's daughters and sons also perform some gender-based agricultural activities like rice winnowing or seed preserving as a part of the family-farming process. Joardar and Kafil, other small farmers of my focus group discussion, help one another through informal labor exchanges.

There are fewer employment opportunities for kamlas today than there once were and many are facing increasing periods of unemployment. Many kamlas want job security and seek a yearly contract locally called *rakhal.* The rakhal works for 365 days a year and performs traditional agricultural tasks from early morning to late night. A rakhal stays in an employer's residence after completing his assigned tasks a day. During their employment contracts, employees like Billal, who work as a rakhal for Tanvir, are allowed to eat seasonal fruits like mangos from the employers' fruit gardens and onions or pumpkins from the fields. Parvin, Billal's wife, reports receiving vegetables, fish and lentils in addition to some cooked food free of costs from Tanvir's wife, Sohali.

One focus group respondent, Keramot, informed me that local ecological resource degradations have led to a reduction in traditional employment opportunities such as the manufacture of traditional tools. Wooden plows, for instance, were traditionally made by day laborers from a local wood known as *langol.* Day laborers like Keramot would also build irrigation infrastructures with local ecological materials such as wood, bamboo, stick and rope. During the rainy season, boats are an important means of transportation and are made with local ecological resources. When field paddy is brought home, it must be husked and winnowed and the winnowing is done with a fan, locally called *kula,* that is made with bamboo and cane. Billal performs this winnowing task as a day laborer of Tanvir. Parvin works in Tanvir's home under his wife Sohali's supervision to winnow paddy, boil it and make rice. The *deki,* or husking pedal, is also made with two pieces of wood and manufactured locally. These agricultural production materials are less used today, partly because the local ecological resources are no longer available. Employment opportunities are thus lost both in their construction and use.

Kafil, Harun and Suman pointed out a number of other ways in which reduction of the Ganges flow decreases traditional employment opportunities in the areas of seed production, plowing, traditional irrigation, livestock caring and fertilizer preparation. Traditional plowing systems are no longer a major source of local employment. Traditional irrigation water supplies have been replaced with the GK project water and deep and shallow tube-wells so construction and maintenance of traditional systems is no longer a source of employment. Bullock carts are being replaced with mechanized vans and trucks. Traditional paddy husking is increasingly mechanized, reducing gender-based job opportunities for Kusum, Morjina and Parvin. Women laborers are not hired to collect drinking water from local water bodies, since that water is no longer available; agricultural laborers are also not asked to prepare fertilizers from traditional sources, since they are no longer available and well-off farmers now purchase chemical fertilizers.

The Ganges flow reduction generates challenges for other occupational groups, like boatmen and fishermen, as well. Currently, many of them are losing their ancestral occupations and are having to leave Chapra. The boatmen cannot transport field crops due to the river flow reduction. Boat-making materials like wood, bamboo and cane are not available in sufficient quantities. Thatchers and basket-makers fail to get employment opportunities due to reductions of necessary local materials. Fishermen cannot catch sufficient fish due to the condition of the Gorai River and local water bodies. Blacksmiths have lost job opportunities due to the imported technologies that now dominate agricultural production. Potters are unable to gather sufficient local materials to carry on their occupation and compete successfully with alternative products.

Traditional ecological knowledge is diminishing in importance as a result of changes to traditional employment and occupational patterns. Previously, the younger people began to learn this knowledge in their early childhood, when they observed their elders' ecological knowledge practices. Grandparents were passionate teachers of their grandchildren. Many children carried breakfast or lunch to their fathers or grandfathers in the fields and observed agricultural practices. Many of them worked as assistants while learning this knowledge. Through this process, Billal developed agricultural production skills by the age of 20. However, these learning practices are no longer important due to the Ganges flow reduction and ecosystem failures. Local ecological knowledge in this setting is not being lost simply through its displacement by alternative scientific and technological knowledge; it is being lost as part of the wholesale transformation of the environment itself and the commodification of common property resources.

New employment patterns

Eighty-five percent of my household survey respondents reported changes to their traditional employment practices, such as agricultural laborers, thatcher, boatmen and potters. The government failed to recognize any of these employment groups as a specialized group. Consequently, these traditional employment systems are in decline because of a government supported "strategy to withdraw laborers from low productivity to higher productivity activities in manufacturing and modern services" (International Monetary Fund 2013:60). Employment sectors are not categorized as "higher productivity" on the basis of wages or working and living conditions, however, so workers who make this transition are not necessarily any better off than before. Many of those who can no longer find employment in traditional rural occupations now work as wage laborers on poultry farms and fisheries projects owned by their more affluent neighbors. Those who work in these new employment sectors do not receive the same benefits as in the traditional system, when they would be fed by the employer or be allowed to pick fruits or vegetables for personal use from the employer's land. Neither do they receive the benefits that should be available to them in a modernized economy, such as sick benefits and weekends off. Jobbar, Fajal and Kusum reported that farm owners control wage and working hours and employees fail to protest unfair working conditions out of fear of losing their jobs. The "new" rural worker thus faces a triple vulnerability: loss of the traditional ecological knowledge; limited skill training for new jobs; and new and more intense forms of workplace uncertainty and exploitation.

Billal stated that the new contract based employment opportunities are not sufficient for him to maintain a decent livelihood. During one conversation, he recounted a glorious past when a day laborer earned sufficient money to maintain a decent livelihood. A day laborer in the 1970s earned 50 *paisas*[3] a day and could save 10 paisas after spending enough for daily necessities like rice, oil and cloth. Other items like fish and wild vegetables were freely available.

As with all the other changes described in this chapter, changes in employment patterns favor the rich over the poor since the rich have ready access to new agricultural technologies and to government subsidies to assist in their purchase. Labor costs for indigenous and HYV crops are much the same (Figure 4.9) but government subsidies for HYV seeds, chemical fertilizers and pesticides, tube-wells and other technology provide powerful incentives for well-off farmers to switch to HYV crop production. The HYV cropping pattern comes together with a reduced need for traditional agricultural labor and the ecological knowledge associated with it.

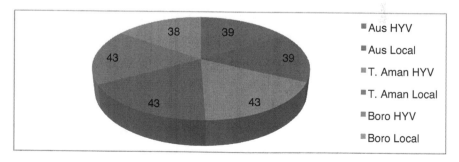

Figure 4.9 Labor costs (US$) for 1 acre of cropland cultivation in 2005–06

Source: Department of Agricultural Extension n.d.

During my focus group meeting in 2011, Tanvir informed me that he has two tractors and two plows, two vehicles, one boat and nine paddy husking machines. This equipment helps Tanvir reduce his labor costs and he can earn extra money by renting them to neighbors. Rahim, Fajal and Laily pointed out a tractor can do the work of 20 laborers. One small van can do the crop transportation tasks of 10 day laborers. One shallow tube-well can provide as much water as 30 day laborers' tasks.

The differential access to training and skill development by different classes also worsens the divide between rich and marginalized farmers. The rich farmers are better prepared to benefit from training programs because of their educational backgrounds. Tanvir completed grade 12 and Tofajjal, an intermediate farmer, completed grade 10. However, Joardar and Billal do not have sufficient reading and writing skills to participate in the training programs that are offered. Joardar dropped out in the ninth grade and has rarely used his literacy skills since then. Billal never enrolled in a school. Elite domination at the local level also ensures that the programs are not designed to accommodate those with less education. Most poor farmers cannot read English and are also disadvantaged as laborers since they cannot read the labels of agricultural materials and technologies. Billal writes Bengali names like *sada sar* (white fertilizer) on containers of urea so that he can perform his job correctly. Sometimes, he makes mistakes and faces scolding and a wage reduction from his employer.

The new agricultural economy creates benefits for those employed at the managerial level. Tanvir, for instance, employs Kafil, an intermediate farmer, as his manager. Kafil's role has expanded in the new economy; he maintains farm machinery and operating schedules, manages irrigation systems, hires and supervises laborers, keeps track of labor costs and sales records and collects crop price updates to assist the marketing process. Day laborers like Billal occupy the bottom layer in this new division of labor and find the gap between them and the rich farmers has widened significantly.

Conclusion

The top-down agriculture policies in Bangladesh are transforming water and agricultural development from traditional to technological systems. Currently, the basin communities at Chapra face major challenges to practice the *kharif*-2, *robi* and *kharif*-1 crop production in coordination with the rainy, winter and summer seasons respectively. The government introduced the different technologies like deep tubewells, HYV seeds, chemical fertilizers and pesticides which displace local knowledge and describe new relationships between water, croplands and communities at Chapra. Consequently, crop production describes major costly practices because Chapra farmers need to spend for fertilizers, seeds, irrigations, pesticides, plowings, laborers and croplands. The reason behind this commodification is that ecosystem-based fertilizers like siltation has replaced with chemical fertilizers and pesticides. As this chapter demonstrates, the process of transformation now underway at Chapra extends into every domain of community livelihoods. Water commodification under conditions of increasing scarcity is central to the transformation but it is also supported by a broad set of agricultural development programs initiated by the government of Bangladesh with the support and the advice of powerful international agencies, upon which the country has grown increasingly dependent over the past several decades. The rich farmers are getting richer and connecting more with globalized economic and concomitant production systems. On the other hand, the vast majority of the population is becoming increasingly marginalized, facing ever more severe livelihood challenges and many are being displaced from their ancestral homes. Many formerly middle-class farmers are either becoming rich or falling into the marginalized class. The central government is not entirely unaware of these problems and has introduced safety-net programs in an effort to reduce displacements and assist marginalized households. Local governments are major implementing agencies of these programs which will be discussed in the next chapter.

Notes

1 Conversion to US dollars was calculated using the 2006 conversion rate. $US1 = 67.4504 Bangladesh Taka (BDT) on July 14, 2006.
2 US$1= BDT 73.1568, July 14, 2011.
3 1 BDT is composed of 100 paisas.

References

Bangladesh Bureau of Statistics (BBS) (National Accounting Wing) 2005. *Table 4.14: Import of Pesticides by Types during 1980–81 to 2000–01*. Dhaka: Government of Bangladesh.

Baviskar, A. ed. 2007. *Waterscapes: The Cultural Politics of Natural Resource*. Delhi: Permanent Black.

Blair, H. W. 1978. Rural Development, Class Structure and Bureaucracy in Bangladesh. *World Development* 6(1): 65–82.

Brandes, O. M. 2005. At a Watershed: Ecological Governance and Sustainable Water Management in Canada. *Journal of Environmental Law and Practice* 16(1): 79–97.

Chowdhury, M. A. T. and H. Zulfikar 2001. *An Agricultural Statistical Profile of Bangladesh, 1947–1999*. Working Paper 54. Bogor: CGPRT Centre.

Cleveland, D. A. and S. C. Murray 1997. The World's Crop Genetic Resources and the Rights of Indigenous Farmers. *Current Anthropology* 38(4): 477–515.

Department of Agricultural Extension (DAE) 2007. *Table 4.09a: Production of Improved Seed by DAE, 1999–00 to 2005–06*. Dhaka: Government of Bangladesh.
——— N.d. *Table 10.4: Per Acre Credit Norm of Some Major Crops for 2005–2006*. Dhaka: Government of Bangladesh.

Escobar, A. 1996. Construction Nature: Elements for a Post-structuralist Political Ecology. *Futures* 28(4): 325–343.

Faber, D. and D. McCarthy 2003. Neo-liberalism, Globalization and the Struggle for Ecological Democracy: Linking Sustainability and Environmental Justice. In

Just Sustainabilities: Development in an Unequal World. J. Agyemen, R. D. Bullard and B. Evans eds. Pp. 38–64. Cambridge, MA: MIT Press.

Guha, R. ed. 2000. *Environmentalism: A Global History.* New York: Longman World History.

Hanson, L. L. 2007. Environmental Justice across the Rural Canadian Prairies: Agricultural Restructuring, Seed Production and the Farm Crisis. *Local Environment* 12(6): 599–611.

Hossain, M. 1988. *Nature and Impact of the Green Revolution in Bangladesh: Research Report 67.* Pp. 1–152. International Food Policy Research Institute.

International Monetary Fund (IMF) 2013. *Bangladesh: Poverty Reduction Strategy Paper (PRSP).* Washington: IMF.

Islam, G. M., P. M. Thompson and P. Sultana 2004a. Lessons and Experience in Inland Fisheries Management and the Impact on Consumption of Fish. *Proceedings of the Workshop on Alleviating Malnutrition through Agriculture in Bangladesh: Bio Fortification and Diversification as Sustainable Solutions.* Dhaka, April 22–24, 2002.

Islam, T. and P. Atkins 2007. Indigenous Floating Cultivation: A Sustainable Agricultural Practice in the Wetlands of Bangladesh. *Development in Practice* 17(1): 130–136.

Khan, A. R. 2013. Indigenous Species of Fish Die. http://arrkhan.blogspot.ca/2013/05/indigenous-species-of-fish-die.html, accessed June 3, 2013.

Kottak, C. P. 1999. The New Ecological Anthropology. *American Anthropologist* 101(1): 23–35.

Lansing, J. S. 2006. *Perfect Order: Recognizing Complexity in Bali.* Princeton, NJ: Princeton University Press.

Latour, B. 2004. *Politics of Nature: How to Bring the Sciences into Democracy.* C. Porter, trans. Cambridge and London: Harvard University Press.

Mascarenhas, M. 2007. Where the Waters Divide: First Nations, Tainted Water and Environmental Justice in Canada. *Local Environment* 12(6): 565–577.

Mehta, L. 2001. The Manufacture of Popular Perceptions of Scarcity in Gujarat, India: Dams and Water Related Narratives. *World Development* 29(12): 2025–2041.

Ministry of Agriculture (MoA) 2007. *Table 4.11: List of Registered Varieties of Notified Crops by Year. Government of Bangladesh.* Dhaka: Government of Bangladesh.

—— 2012. *Free Seed Distribution for Monsoon Seasonal Paddy Production.* Dhaka: Government of Bangladesh.

Ministry of Finance (MoF) 2012. *Safety Net Programmes.* Dhaka: Government of Bangladesh.

—— 2013. *Safety Net Programmes.* Dhaka: Government of Bangladesh.

Paul, B. K. 1984. Perception of Farmers and Agricultural Adjustment to Floods in Jamuna Floodplain, Bangladesh. *Human Ecology* 12(1): 3–19.

ProthomAlo 2013b. Fisheries Project with Embankment over the Dying Padma River. http://archive.prothom-alo.com/detail/date/2013–02–18/news/330047, accessed May 12, 2014.

Rahman, M. H. 2012. Livestock. http://www.banglapedia.org/HT/L_0133.htm, accessed January 3, 2013.

Robeyns, I. 2005. The Capability Approach: A Theoretical Survey. *Journal of Human Development* 6(1): 93–114.

Scott, J. C. 1998. *Seeing like A State: How Certain Schemes to Improve the Human Condition Have Failed.* New Haven and London: Yale University Press.

Sillitoe, P. 1998. What, Know Natives? Local Knowledge in Development. *Social Anthropology* 6(2): 203–220.

Water Resources Planning Organization (WARPO) 2009. *Annual Report (July 2008–June 2009): Water Resources Planning Organization.* Dhaka: Government of Bangladesh.

Zaman, M. Q. 1993. Rivers of Life – Living with Floods in Bangladesh. *Asian Survey* 33(10): 985–996.

5 Local governments' roles to protect the marginalized communities

Local-level government in rural Bangladesh administers a number of agricultural subsidy and safety-net programs established by the national government on the basis of guidelines provided by the Poverty Reduction Strategy Paper and the Millennium Development Goals. In theory, these programs could alleviate some of the worst hardships caused by the ecosystem and crop production failures described in previous chapters. Since many components of the water management system, such as the Farakka diversion in India and the GK Project in Bangladesh, are entirely outside the control of communities like Chapra, these communities have become increasingly dependent on local governments to manage the scarce and unpredictable water supply that remains. Communities expect local-level government to ensure agricultural production and livelihoods and, in order to do so, it needs to address seed and fertilizer management, as well as water management and provision of social service benefits pertaining to food, employment, education, health care and housing programs. Such programs have, in fact, been implemented but, as is so often the case in this setting, they fail to incorporate the voices of the marginalized communities and have fallen well short of meeting their goals. In this chapter, I critically describe the performance of these programs in the context of local power structures. My analyses suggest that local elites in Bangladesh, based on their connections to the national government and external agencies, are able to manipulate subsidy and social programs and thereby extend their resource control over rural communities.

Power structure of local governments

Local elites are major decision-makers in local government institutions that implement central government decisions about agricultural policies and safety-net programs. As outlined in chapter two, the political system in Bangladesh is highly centralized and local-level government institutions at the Upazila Pasrishad (UP) and Union Council (UC) levels are controlled by local elites acting in concert with national elites, including elected MPs, high-level bureaucrats and high-ranking officials of the political party in power. All UPs have similar administrative departments, for example, the Department of Agricultural Extension

(DAE) and the Local Government Engineering Department (LGED), which constitute the "street level bureaucracy" (Heyman 2004:493) within this system. The DAE and LGED are local offices of national departments run partly by the UP. Innovation occurs almost entirely in the form of what Hayami (1981:169) calls "induced innovation," that is, as a consequence of the actions of external elites in coordination with local elites, rather than as local grass-root responses to local or regional issues.

UPs and UCs also manage local social service agencies that deal with issues like water management, seed and fertilizer distribution and subsidies and food, employment, education and health care programs. The Upazila Nirbahi Officer (UNO) is the chief administrator who serves as a coordinator among UP and UC departments and committees, serving as the chair or sub-chair of the committees that administer the above services.

Every elected member of a UC is a rich farmer from a local village. These local elites are, thus, the major power brokers between the marginalized people and the ruling elites who promote the central government's control over local resources. Similar situations have been described by Agrawal and Ostrom (2001) in the context of their study of the World Bank's decentralization approach in India. The UP is also dominated by local elites but, as noted in chapter two, many UP resource management committee members are appointed rather than elected. Local elites are able to exploit local power structures to ensure their control of UCs and UPs. Tanvir, Tofajjal, Joardar and Billal provided me with details about how this exploitation works at Chapra. Joardar, a small farmer with minimal landholdings, works for Tanvir, a wealthy farmer, needs access to safety-net programs like health care or housing and government support to obtain seeds and fertilizer. Joardar also needs to rent land from Tanvir on occasion, for sharecropping. As a sharecropper, he needs to vote in local elections according to Tanvir's preferences and failure to do so can lead to the loss of sharecropping land or government benefits. Eighty-three percent of my household survey respondents report similar forms of exploitation by local elites. Elites also do not keep election promises to poor farmers, such as the inclusion of Joardar on local resource management committees.

Local Members of Parliament (MPs) also exercise extreme levels of control over power structures at the local government level in ways that are not conducive to democratic practices or inclusion of marginalized people. The MP controls local resource management like irrigation water, seed, food, education, employment, health care and housing by controlling the nominations of chairmen and member candidates for UP and UC elections. Many local elites (all wealthy landowners) practice nepotism or bribery to get these nominations (New Age 2014). They consider these bribes as investments that make profits for them later through local resource management decisions (Khan 2013). This system effectively excludes marginalized people from receiving a nomination to run for UC or UP positions. This top-down elite domination corresponds to what Scott (1998:6) has termed the implementation of a "high-modernist" agenda. Ferguson similarly (1994:15) argues that "a rural development project is a part of the expansion of the capitalist mode of production, which is often not so good at all for the poor peasants." Local government involvement in this type of capitalist expansion leads to what many researchers have termed the destruction of "local commons" (Bardhan 1993) and "sovereign selves" (Agrawal 2003).

Water management

The central government controls the Gorai River Restoration Project and Ganges-Kobodak Project described in chapter two, while local government manages smaller scale water projects, like the Shallow Tube-Well and Deep Tube-Well Projects, created as part of the country's Poverty Reduction Strategy (International Monetary Fund 2013:283). The large projects are thus controlled by national elites, while the local activities are controlled by local elites. The average community member has no opportunity to raise their voice in the decision-making process for any of these projects.

Local activities are managed by water management committees at the district and sub-district (upazila) levels (Table 5.1) created on the basis of guidelines provided by the central government in Bangladesh. At the upazila level, the UP chairperson and the UNO are the chair and vice-chair respectively of the irriga-tion management committee. Ten of the remaining 11 positions are filled by government staff from various other agencies and only one position is filled by a representative of the more than 200,000 farmers in the Kumarkhali Sub-district. This representative farmer is selected from among the local elites and works as a middleman between local governments and the other farmers in the sub-district. Joardar termed this representative as *sorkarer dalal* (an agent of the government).

Table 5.1 clearly demonstrates that there is virtually no involvement from community members in local government irrigation management committees. The upazila committee simply implements the central government's irrigation policy guidelines. This committee is officially responsible for coordinating with NGOs based on these guidelines. There are no irrigation committees at the union council level, even though unions and villages are the major focusing points for these irrigation management activities of the committees.

None of the water management systems in the Chapra region are being effectively operated. As documented in chapter two, the GRRP is producing ecological vulnerabilities like flooding and river bank erosion that the local upazila irrigation management committee does not have the authority to correct even if they did have the political will. The GK authority is unable to provide sufficient water supplies due to reduced flows in the Ganges River and has not coordinated their supply system with the agro-ecological system that includes high, middle and low elevation croplands. As focus group respondents, Kamal and Kafil, informed me, the higher croplands face drought when the project authority provides proper water supplies to lower croplands; whereas, lower croplands face water pooling and stagnation when they provide adequate water

Table 5.1 Irrigation management committees at district and upazila levels

District Level Members

Deputy Commissioner (Chair)

Executive Engineer, Bangladesh Agriculture Development Corporation or Senior Agriculture Engineer, Department of Agriculture Extension (Secretary)

Superintendent of Police

Deputy Director, Department of Agriculture Extension

Deputy Director, Bangladesh Rural Development Board

Representative of the Department of Environment

Executive Engineer, Local Government Engineering Department

Representative, Health Department Executive Engineer, Bangladesh Water Development Board

Executive Engineer, Power Development Board

Executive Engineer, Rural Electrification Board

Representative, Fisheries Department

Sub-District or Upazila Level Members

Upazila Parishad chairman (Chair)

Assistant Engineer, Bangladesh Agricultural Development Corporation or Department of Agricultural Extension (Secretary)

Upazila Nirbahi Officer (Vice-Chair)

Upazila Agriculture officer, Department of Agricultural Extension

Upazila Engineer, Local Government Engineering Department

Officer, Upazila Rural Development

Representative, Bangladesh Water Development Board

Representative, Health Department

Officer, Upazila Fisheries Department

Representative, Power Development Board/Rural Electrification Board

Officer in Charge (OC), Police

Representative Farmer

Source: Ministry of Agriculture (2011)

supplies to the higher croplands. The central government hoped to overcome these problems by establishing tube-well projects which, though under the control of the richer farmers, have the benefit of localizing the water delivery system. However, these projects are currently facing major challenges as well. Sorkar, Taher and Kamal informed me the deep tube-wells withdraw arsenic contaminated groundwater and that water levels are dropping. The groundwater depth was 7 meters in 1981 but that had decreased to 3.5 meters by 2010 (Ministry of Finance 2013:61). The upazila irrigation management committee is fully aware of these challenges but the centralized approach of the national government does not provide them with the means to find remedies.

The central government has recognized and attempted to resolve water drainage problems throughout the country, including Chapra where portions of the Kaligangya River were cut off by GK Project canals and portions of other local water bodies were cut off by constructions of roads and highways. The central government provided $79,994,500, $170,685,000 and $344,386,000 in 2009, 2010 and 2011 respectively, to overcome water drainage problems throughout Bangladesh (Ministry of Finance 2012). Local government officials and the irrigation management committees control whatever funds come to their region and, once again, project managers fail to consult with community members about these problems. If marginalized people resist the actions of the local committee they are reported to the police as has having carried out illegal activities against government rules and regulations. It is for this purpose that the local irrigation management committee always includes a representative of the local Police Office. This committee structure is a major reflection of the local power structure that benefits the rich people and deprives the marginalized people.

Seed management

Local elite control of resource management also extends to seed management. As described in chapter four, most farming households find it difficult to produce their own seeds due to lowered crop productivity and the government is actively discouraging the use of local seeds in favor of HYV seeds. Committees are formed at district, upazila and union council levels on the basis of central government guidelines and these committees organize the distribution of HYV seeds and chemical fertilizers and, through safety-net programs, give a small amount of seed to farmers who cannot afford to buy it. The UP committees determine who will be licensed to sell seeds and they must submit progress reports periodically to the central government. As in the case of the water management committees, very few local farmers are able to serve on these committees which are dominated by appointed civil servants (Table 5.2). Local elites compete for membership on these committees and use them to promote their own interests.

Joardar and Billal argued that local elites control the seed business by acquiring the license to sell the seeds. Local governments collect the HYV seeds from the central government and provide them to local dealers. There are 54 seed dealers in Kushtia and three of them are in Kumarkhali Upazila where Chapra is located (Ministry of Agriculture 2009b). The seed dealers control seed prices and the poorer farmers like Joardar face major challenges for getting seeds of proper quality and price in time for planting. According to Joardar, the committee chairman and its members dismiss those who complain as supporters of the opposition political party who want to tarnish the ruling government's reputation.

Table 5.2 Seed and fertilizer management committees in local governments

Sub-District or Upazila Level Members

Member of Parliament (Advisor)

Upazila Parishad Chairman (Advisor)

Upazila Parishad Vice-Chairmen (Advisor)

Upazila Nirbahi Officer (Chair)

Upazila Agricultural Officer (Secretary)

Officer, Upazila Animal Resource

Officer, Upazila Fisheries Department

Officer, Rural Development

Officer, Cooperative Department

Officer in Charge, Police

Union Council chairmen

Representative, Bangladesh Agricultural Development Corporation

Representative, Bangladesh Fertilizer Association

Farmer, nominated by Upazila Parishad

President, Upazila Press Club

Union Level Members

Union Council Chairman (Chair)

Union Council Secretary (Secretary)

Union Council Members

Two respectable local persons, Member of Parliament nominated

Local *Imam* or spiritual leader, Member of Parliament nominated

Representative, Teacher, Member of Parliament nominated

Officer, Deputy Assistant Agriculture Officer Upazila Nirbahi Officer nominated

Source: Ministry of Agriculture (2009a)

Fajal, Faruk, Mofiz and Kusum informed me that the poorest farmers cannot afford the commercial HYV seeds and that the central government has created safety-net programs to assist them. The central government provides guidelines for implementing these programs through agricultural rehabilitation committees created at the district, upazila and union levels (Table 5.3). The district committee finalizes the total amount of seed grants for each upazila and distributes them to all upazila within the district. The upazila committees in turn distribute them to all union committees within the upazila. The union committee finalizes a list of the total vulnerable people in the union, then distributes the free or subsidized seeds to the listed people. During the period of my fieldwork, the government was distributing HYV seeds to poor households under two major programs: one for rice, wheat and maize and the other for pulse and oil seed.

Table 5.3 Agriculture rehabilitation committees in local governments

Sub-district or Upazila Level Members

Upazila Nirbahi Officer (Chair)

Officer, Upazila Agricultural Officer (Secretary)

Officer, Upazila Animal Resources

Officer, Fisheries Department

Officer, Rural Development

Officer, Project Development

Representative, Joint Task Force

Representative, Non-Government Officer

Union Level Members

Union Council Chairman (Chair)

Senior Deputy Assistant Agriculture Officers (Secretary)

Union Council Members

Representative, Non-Government Organization

Deputy Assistant Agriculture Officers

Source: Ministry of Agriculture (2009a)

In 2012, the records of the Ministry of Agriculture reported they provided free kharif-1 season rice seeds of *Bona Aus* (Nerica) and *Ufsi Aus* for 1,500 acres of cropland in Kushtia (Ministry of Agriculture 2012). The government gives this seed to local farmers free as a major initiative to popularize this HYV seed. This allocation is only enough for 3,000 of the 215,768 marginalized farmers in Kushtia District who own less than 2.49 acres of croplands (Bangladesh Bureau of Statistics 2005). Furthermore, the government only provides this seed for the kharif-1 cropping period and not for two other longer periods of kharif-2 and robi.

All farming households that own less than 2.5 acres of cropland are eligible to get a maximum of one-third of an acre of cropland seed. The government has specific guidelines for local decision-making processes of distributing this free seed. The district agriculture rehabilitation committee finalizes a priority list of the marginalized farmers submitted by the upazila committees. After receiving the approved list from the district, the upazila agricultural rehabilitation committee directly distributes the seeds to the listed farmers. Joardar and Billal insisted that the Union Council committee chairman and members include their supporters on the list of recipients and exclude those who support their political opponents.

Raju, Sabrina and Bakar informed me that many students of Kushtia Government College, who are the ruling government supporters, received the free rice seeds in 2011 although they do not have any involvement in agricultural production. According to them, nobody gets these agricultural benefits who fail to "satisfy" the officials of UC or UP committees. Local elites also exploit the marginalized farmers by other mechanisms, for instance, by extracting free labor. Tanvir exploited Malek's labor on his own farm in exchange for the government's free seeds. This elite exploitation of local seed programs thus increases marginalized farmers' seed deprivation and agricultural production challenges.

Fertilizer management

As noted in chapter four, farming households traditionally rely on siltation, algae, earth-worms, cow-dung and water hyacinth as natural fertilizers but are unable to obtain these materials in sufficient quantities today, due to ecosystem failures caused by government interventions and the reduced flow of water in the Ganges Basin. They therefore need chemical fertilizers to maintain crop productivity and the national government has implemented fertilizer distribution programs for this purpose. The fertilizer distribution system is managed by the same committees that oversee seed distribution and with all the similar outcomes (Table 5.1). Members of the local elite monopolize the sale of fertilizers at prices that poorer farmers cannot afford and manipulate the distribution of free seeds so as to consolidate their authority over Chapra villagers and political supporters. Kofil, Tanvir's son, is a fertilizer dealer in the Kumarkhali upazila. He augments the amount of fertilizer he can sell by obtaining a retail license both for himself and for an appointed salaried worker, Bokul.

Most smallholder and all marginalized farmers cannot afford to buy chemical fertilizers at full market price. The cost of urea fertilizer, for instance, an essential source of nitrogen, was Bangladesh Taka (BDT) 12 per kilogram in 2009, increased to BDT 20 (US$0.26) in 2011 (Ministry of Finance 2013:68). This 40 percent increase means that one farmer needs to spend an extra US$40 for 1 acre of cropland in a year. Poor farmers therefore desperately need access to this small amount of free fertilizer that the government distributes. In 2012, the Ministry of Agriculture provided free fertilizer in Kushtia for 1,500 acres of cropland to the farmer owning less than 2.5 acres of cropland being eligible to receive 16 kilograms of urea, 10 kilograms of Diammonium Phosphate (DAP) and 10 kilograms of Muriate of Potash (MOP) to produce the *Ufsi aus* crops and, 16 kilograms of urea, 10 kilograms of DAP and 10 kilograms of MOP to produce the *bona aus* crop (Ministry of Agriculture 2012). The total market value of this free fertilizer is US$19 but 1 acre of cropland, on average, requires US$130 of fertilizer per year. Farmers desperately need more than the fertilizer the government provides for free. Smallholders who do have some cash assets will thus buy as much additional fertilizer as they can, even at the inflated prices sometimes created by local price-fixing. According to them, local governments ignore their complaints regarding price manipulation due to the nexus between the local government officials and fertilizer dealers. According to Jobbar and Harun, as with complaints about seed distribution, government officials argue that only opposition political party supporters lodge these complaints to tarnish the image of a ruling government. Consequently, the marginalized people face double vulnerabilities: political repression and agricultural production challenges. In summary, this fertilizer management system promotes elite interests and deprives marginalized people of key agricultural resources.

Food security program management

Local governments' water, seed and fertilizer management failures inevitably result in major food insecurity for the marginalized people. In order to address this problem, the central government created the food safety-net programs to secure minimum food rights for them, based on the guidelines of Poverty Reduction Strategy Paper (Ministry of Food and Disaster Management 2010) and the Millennium Development Goals (Ministry of Food 2013:19). As in the cases described above, the central government places local governments in charge of implementing these programs.

As with the water and seed and fertilizer committees, the local government food management committees operate within a local power structure that excludes marginalized people (Table 5.4). Committees at the district, upazila and union levels operate several food programs such as the Vulnerable Group Feeding and Food For Work programs. Some major activities of the district food management committee are to review previous project performances, allocate money to upazila committees for the current program and review project activities. The upazila food committee approves the priority lists of the target people that are submitted by the union council food committees. Based on this approval process, the union council food committee distributes food benefits to eligible people.

Fazlu, Abonti and Bakar explained to me some of the ways that food programs are used to promote elite interests. Local MPs control the district and upazila food management committees based on their advisory positions. As a result of their political control, the district food committee members are local political leaders who sometimes violate the central government's own guidelines (Raihan 2013). Local elites at Chapra, Tanvir, Mokles and Khalek, are therefore key decision-makers when it comes to developing a list of eligible food benefit recipients. According to my informants, these lists are drawn up on the basis of personal interests, nepotism, bribery and political bias, rather than actual need.

One of the most important and most funded food relief programs is the Vulnerable Group Feeding (VGF) program. This program had a budget of US$700,000,000 in 2012–13 and provided food to 400,000 people. People who own less than 0.15 acres of cropland are eligible to receive 10–20 kilograms of rice or wheat a month.

Local governments provide specific benefits to poor households during special occasions like *Eid-ul-Azha*, an important Islamic holiday. In 2010, the VGF program provided a special food ration to 3.5 million people in 482 upazila for the *Eid-ul-Azha* celebration. Every VGF card holder received 10 kilograms of rice (Ministry of Disaster Management and Relief 2010). The VGF program

Table 5.4 Food For Work or KABIKHA committees in local government

Sub-district Level Members

Member of Parliament (Advisor)

Upazila Parishad Chairman (Chair)

Officer, Project Implementation Management (Secretary)

Upazila Nirbahi Officer (Vice-chair)

Upazila Parishad Vice-Chairmen

Executive Engineer, Local Government Engineering Department

Officer, Upazila Agricultural Department

Officer, Upazila Rural Development

Officer, Upazila Social Welfare

Officer, Upazila Account Department

Officer, Upazila Fisheries Department

Comptroller, Upazila Food Department

Officer, Upazila Assistant Engineer, Local Government Engineering Department

Officer, Upazila Youth Development

Union Council Chairmen

Two respectable persons, Upazila Nirbahi Officer nominated

One teacher, Upazila Nirbahi Officer nominated

One woman, Upazila Nirbahi Officer nominated

Union Level Members

Union Council Chairman (Chair)

Union Council Secretary (Secretary)

Union Council Members

Deputy Assistant Agriculture Officer

Member, Family Planning Visitor

Field Assistant, Bangladesh Rural Development Board

Three respectable persons, Upazila Nirbahi Officer nominated

One teacher, Upazila Nirbahi Officer nominated

One woman, Upazila Nirbahi Officer nominated

Source: Ministry of Food and Disaster Management (2012a)

also provides free food during *Eid-ul-Fitr*, another religious holiday that occurs at the end of Ramadan. These VGF allocations ensure the participation of marginalized households in important Muslim religious celebrations, as well as provide them with a basic level of nutrition throughout the year. However, this program is not provided to non-Muslims during religious holidays and the program is discriminatory on that basis. The political motivations and economic

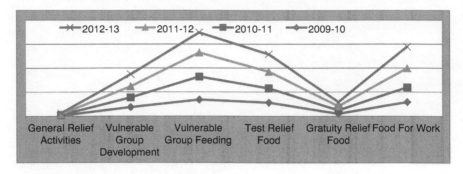

Figure 5.1 The central government's food safety program budgets (US$[1])

Source: Ministry of Finance (2012)

1 US$1=BDT 80.3028, June 6, 2012

self-interest of those who control the food committees also means that many eligible recipients of food do not receive it. Mofiz, for instance, failed to get the VGF benefits in 2010 because of local elite domination over the VGF program and he sent his wife and children to his father-in-law's house to celebrate the *Eid-ul-Azha* holiday. Abonti took an NGO loan to buy special food for his children during the *Eid-ul-Fitr*. This exclusion and multiple socioeconomic challenges are an every-year experience for marginalized people at Chapra.

The central government has also funded a Food for Work (FFW) program, a 60-day program that generates employment opportunities for marginalized people. One employed person, based on this program, is eligible to receive 8 kilograms of rice or wheat a day as payment for providing labor for the construction or maintenance of local infrastructure, such as GK project canals, embankments and roads (Ministry of Finance 2012). However, the FFW program goals are thwarted by the local power structures and FFW program implementation lacks transparency and accountability. Based on Bangladesh Bureau of Statistics (2005) data, 41 percent of people in Kushtia who own less than half an acre of cropland report they did not receive FFW program benefits in 2010. Although the safety-net programs have the capacity to mitigate food shortages, they are mismanaged and do not operate on a scale sufficient to alleviate more than a small percentage of the food shortages being experienced in Chapra and throughout Bangladesh. As I will argue in the next chapter, since these food shortages are entirely the outcome of human interventions, they constitute an abuse of the human rights of smallholder farmers and agricultural laborers in this setting.

Employment program management

As documented in chapter four, employment patterns at Chapra have changed radically over the past several decades. New jobs have been created in sectors like fish farming, tube-well construction and maintenance, and operation of the new farm machineries used for paddy husking and plowing. These new jobs are far fewer in number than the jobs lost in traditional fisheries, conventional irrigation system, maintenance and the practice of traditional farming techniques that include the collection of many freely available natural resources for use as fertilizer, for the construction of tools or for sale in local markets. To mitigate these losses local governments have created various employment management committees at the district, upazila and union levels based on guidelines provided by the central government.

Local governments manage several employment programs some of which are specialized in the economic sectors they wish to promote. They include: (1) the Employment Generation Program for the Poorest, (2) Fund for Assistance to the Small Farmer and Poultry Farms, (3) *Swanirvar* or Independence Training Program, (4) Employment Generation for the Ultra Poor, (5) Rural Employment for Public Asset, (6) Rural Employment and Rural Maintenance Program, (7) Micro-credit for Women Self-employment and (8) *Jatka* (Fish) Protection and Alternative Employment for Fishermen. The central government provided a total of US$498,115,000 over 4 years from 2009–10 to 2012–13 to promote employment opportunities for 400,000 marginalized people in the poultry industry through the Fund for Assistance to the Small Farmer and Poultry Farms. This program pays for the importation of poultry species that are used to replace domestic species considered not suitable for commercial poultry operations. A second program, the *Swanirvar* program, trains unemployed people in how to set up and operate these poultry businesses.

In this section, I focus mainly on a less specialized form of assistance, the Employment Generation Program for the Poorest (EGPP) (Table 5.5). In addition to the employment management committees at district, upazila and union levels, the EGPP program requires a Project Implementation Committee (PIC) at the bottom level of the committee hierarchy. The UC chairman nominates five to seven UC members to serve on PIC excluding all marginalized people even though they are the programs' target group. The union committee is responsible for EGPP project implementation while PIC looks after everyday activities such as taking attendance and paying the employed people.

Suman, Jobbar, Fazlu and Kamal reported that the ruling elites control the EGPP for their own interests just as they do all the other local committees. Many eligible participants, including Suman and Jobbar, reported not even receiving news about their eligibility for EGPP employment. Fazlu and Kamal also pointed out that there is a huge gap between the total number of unemployed people and the number employed by government programs. In 2010–11, for example, the central government offered EGPP opportunities to 2,380 out of 98,450 eligible people in the Kamarkhali Upazila in Kushtia (Ministry of Food and Disaster Management 2012b). Suman and Jobbar argued that the ruling elites in local governments exploit the gap between government employment opportunities and the total number of unemployed in order, once again, to consolidate their positions within the local power structure.

Jobbar, Fazlu and Mofiz reported that local elites will sometimes pay them from EGPP project funds, rather than own personal funds, for work they

Table 5.5 The EGPP committees at upazila and union levels

Upazila Level Members

Member of Parliament (Head Advisor)

Upazila Parishad Chairman (Advisor)

Upazila Nirbahi Officer (Chair)

Officer, Upazila Project Implementation (Secretary)

Upazila Vice-Chairmen

Union Council Chairmen

Officer, Upazila Health and Family Planning

Officer, Upazila Agricultural Department

Officer, Upazila Fisheries Department

Officer, Upazila Engineer, Local Government Engineering Department

Comptroller, Upazila Food Department

Office, Upazila Education Department

Officer, Upazila Youth Development

Officer, Upazila Women Affairs

Officer, Upazila Cooperative Department

Officer, Upazila Animal Resources

Officer, Upazila Social Welfare

Officer, Upazila Ansar & Village Defence Party

Officer, Upazila Rural Development

Member, Upazila Field Supervisor, EGPP

Teacher, Upazila Nirbahi Officer nominated

Volunteer representative, Upazila Nirbahi Officer nominated

Manager, Upazila level Banks

One respectable person, Upazila Nirbahi Officer nominated

Union Level Members

Union Council Chairman (Chair)

Union Council Secretary (Secretary)

Union Council Members

Union Deputy-Assistant Agriculture Officer

Manager, Local Bank

Field Assistant, Bangladesh Rural Development Board

Women Member of Union Council reserved seats, Upazila Nirbahi Officer nominated

One teacher, Upazila Nirbahi Officer nominated

Source: Ministry of Food and Disaster Management (2012a)

perform for the elite farmers on their farms, work that is not part of the EGPP program. Jobbar and Fazlu, for instance, will be given work at Tanvir's deep tube-well project when they are supposed to work on GK project canal maintenance. They do not protest this malpractice out of fear of losing the EGPP work. They also reported that, many people whose names are on the EGPP benefits list are local political leaders and their family members including brothers, brothers' wives or domestic servants. None of them work on approved EGPP projects but they all receive full payment from EGPP funds. Local elites are also reported to put false names on the EGPP list in order to gain access to additional benefits. Local government authorities know about many of these malpractices but they fail to execute legal action, due to their own limitations within the local power structure.

Alamgir, Bakkar and Faruk argued that the employment programs actually increase their employment challenges. This occurs because the central government prefers to fund projects that displace local practices and further erode their self-sufficiency. The central government provides micro-credit loans to help women establish poultry farms using imported species but focus group respondents Abonti, Gedi, Kusum and Morjina argued that they would like to support for their contribution to domestic species production. The *Swanirvar* program provides skill development for raising foreign poultry species but this displaces local ecological knowledge regarding domestic animal husbandry and creates dependencies on distant markets and vulnerability to market conditions they cannot control. The rural infrastructure maintenance programs help to maintain government infrastructures like the GK project but are not directed towards the restoration of local water bodies that are blocked by GK project infrastructure. My informants also pointed out that the central government's employment programs do not recognize traditional occupational groups like blacksmiths and potters. These top-down employment programs protect technocentric development infrastructures that promote elite interests and exclude marginalized people and subsistence livelihoods.

Furthermore, the bureaucratic system of payment for those who do gain access to these programs also encounters problems. Suman, Alamgir and Kamal, for instance, report having difficulties to open bank accounts in order to be paid by the EGPP program. Many of them have never had bank accounts and do not understand the procedures and requirements for opening an account. Many do not have property documents, for instance, or other forms of documentation necessary to confirm their identity. Suman reported that bank officers treat him as a low-caste person whenever he has bank business, a process that reinforces his sense of marginalization and discourages his participation in government employment programs.

Education program management

The focus group respondents agreed that their ability to provide education to their children is compromised when agricultural production is low or employment opportunities are not available. My informants disagreed, however, on the type of support they wanted from government in the area of education. Barek and Morjina stated that they would prefer government support for agricultural production and employment, so they could pay education costs themselves, rather than have the government create special support programs for the poor. Parents from different socio-economic backgrounds also place very different demands on the public education system. Tanvir, a wealthy landowner, is interested in having his children obtain enough education to maintain the family status in the local power structure. Tofajjal, an intermediate landowner, argued that a good education is a better investment for his children than money spent in the agricultural sector. Joardar, on the other hand, a smallholder, desperately wants education for his children in the hope that they can reverse the pattern of marginalization he has experienced when it comes to participating in local resource management processes. Ninety-four percent of household survey respondents hold the belief that educated children could help them get better access to government programs and services. Billal, a landless laborer, dreams of his second son, Lablu, becoming a government magistrate in order to protect their rights. However, 67 percent of the marginalized household survey respondents reported not being able to afford the costs necessary to educate one or more of their children.

Local governments administer education support programs at the district, upazila and union levels based, as always, on the guidelines of the central government. There is also a School Management Committee for every school that is entrusted with the task of implementing the actual programs.

Parents do not need to pay tuition or school fees for their children in order to attend primary and secondary institutions up to grade 10. However, they need to pay fees for grades 11 and 12 and university which are also highly subsidized. Despite this free education to grade 10, poor households cannot afford other educational expenses such as books and school uniforms. To assist with these costs, the central government has created eight scholarship and grant programs: (1) Stipend for Disabled Students, (2) Grants for Schools of the Disabled, (3) Stipend for Primary Students, (4) School Feeding Program, (5) Stipend For Dropout Students, (6) Stipend For Secondary and Higher Secondary Level Students, (7) Preliminary Education for Development of Children and (8) Post Literacy Education Project (Figure 5.2). These education programs are developed based on policy guidelines of Poverty Reduction Strategy Paper (Ministry of Food and Disaster Management 2010). The government is also decreasing government education budget and encouraging private sector involvement in establishing new schools. In 2012, because of this government education policy, there were 3 government colleges and 10 schools in Kushtia District whereas there were 30 private colleges and 173 private schools (Kushtia Zila Samity 2012).

In addition to the fact that school costs are more than some households can manage, some of my focus group respondents, Billal, Suman, Sabbir and Tarun, indicated that they need to keep children home to help them with basic subsistence tasks. When I asked Billal about access to formal education for his

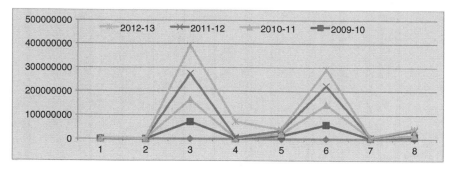

Figure 5.2 The eight scholarship and grant program budgets
Source: Ministry of Finance (2012)

children, he expressed a deep concern with the comment "*bhaat paina, shikhya dibo kibabe?*" (How can we afford education for our children when we do not have enough food for the whole family?) Many poor families are thus forced to make choices between food and education. Suman and Sabbir reported making different choices for their male and female children. Based on their limited economic capability, they prefer to educate a son than a daughter with their logic influenced by *bongsher baati* (line of inheritance[1]). The governments operate school food programs but one focus group respondent, Tarun, informed me that his children were not able to benefit from this much valued program because his children do not have suitable conditions at home to study and he therefore cannot send them to school. He cannot afford a light for his children to study at night and his whole household resides in a single cottage, locally called *kureau ghor*, where his children do not have even a minimum sleeping space. Many children with these economic challenges, fail to attend school on a regular basis, do not complete their grades, and therefore, are not eligible for scholarship benefits.

Many focus group respondents, including Kamal, Rahim and Borkat, stated that scholarship and grant programs are currently under political control. As in the case of other social services, local MPs and their supporters on district, upazila and union councils control the school management committees for their personal interests. Head teachers are dependent on the upazila and district administration for their promotion, salaries and job postings. Based on this dependency, many staff members in schools use their time doing private consultancies and are dependent for their positions on nepotism and political support from local elites. The school management committees are unable to take legal action against these malpractices even when some of their members would prefer to do so. Under the circumstances, many of the program benefits end up going to the children of rich farmers and their relatives, while children from poor households receive nothing and are forced to drop out from school in high numbers. Consequently,

84 percent of my marginalized survey respondents reported not receiving any benefits from the government's Stipend For Secondary and Higher Secondary Level Students. Among these respondents, 50 percent dropped out from school before completion of grade five.

Health care program management

Local governments perform important roles in implementing the central government's health care programs. Unfortunately, these health programs are also controlled by local elites similar to the education programs. Illness is part of the vicious cycle of marginalization for poor households, the result of too little food and poor nutrition, which then, in turn, causes food production and household income to drop still more. When a marginalized household head encounters sickness, the whole household faces economic crises and multiple sicknesses. Fifty-six percent of household survey respondents report encountering sickness in 2010 and needed support from local government to regain their health and well-being. Seventy-two percent of my survey respondents answered yes to the question: "did you encounter agricultural production failures due to sickness? Fifty-two percent reported sickness associated with crop failures during all three cropping periods.

According to Bakar, Borkat, Suman and Sabbir, many of the illnesses they experienced are preventable but preventive health care is not funded by the central government and they cannot afford to pay for curative health services. Currently, local people suffer from high incidences of malnutrition as a result of food shortages and from diarrhea caused by unclean drinking water and water contaminated by chemical fertilizers, insecticides, fungicides and herbicides. Sixty-one percent of survey respondents report health problems in 2010, caused by unsafe drinking sources. Some of these problems could be resolved by government health care programs that provide benefits such as free tube-wells and sanitary latrines but local elites control these resources. Eighty-four percent of my survey respondents report having no access to tube-wells or sanitary facilities. Some foods, notably fish caught in local water bodies subject to agricultural field run-off, are also a source of illness. Because of the overuse of TSP fertilizer on rice crops, one kilogram of rice in Kushtia, on average, has been found to contain 0.099 milligrams of cadmium (Bhuyan 2013). Cadmium, like other heavy metals such as mercury, can create diseases like kidney failure and cancer. As stated earlier in this chapter, deep tube-wells often withdraw arsenic contaminated water and this is also a source of health problems. The Farakka diversion in India can be considered a significant contributing factor for arsenic contamination since the deep tube-wells have been built mainly as a way to make up for the lack of natural flow in the Ganges and Gorai Rivers (Rahman et al. 2010). Currently, about 85 million people in Bangladesh are facing arsenic contamination in drinking water and field crops. Forty-four percent of the total population in Kushtia is facing arsenic contamination (Hossain 2006). Despite the pervasiveness of illnesses caused by environmental factors of this type, the central government's health care services do not include any preventative measures and do not monitor or enforce water quality regulations.

The failure to provide preventive health services dramatically increases the disease burden for poor households at Chapra. Many diseases like cadmium poisoning are new to them and have no remedy within traditional medical systems. To make matters worse, the central government is reducing their expenditures on doctors, nurses and new medical technologies based on the terms of the current structural adjustment program. Community clinics and NGO-operated first aid service centers are being introduced based on the guidelines of the Poverty Reduction Strategy Paper and Millennium Development Goals. This further contributes to the commodification of local health care services. Thirty-five community clinics have now been established in Kushtia (Ministry of Health & Family Welfare 2013) mainly in urban areas. The privatization of health services benefits those who can afford the services but this is not the case for most marginalized people. Eighty percent of my survey respondents report not being able to receive needed health services due to lack of ability to pay. Since the central government understands that many people cannot afford privatized services, they have encouraged more NGO involvement in providing health services. Figure 5.3 depicts the pattern of health service provision in rural Bangladesh as a whole.

Standard health services in Bangladesh are generally organized in four tiers. Division level hospitals and medical colleges serve the several districts located within them;[2] district level primary hospitals serve the district population, an upazila health complex serve the sub-district population and union level health and family welfare centers serve a village or group of villages. The Kumarkhali Upazila Health Complex Bulletin (2012) indicates there are 12 doctors at this facility, serving a population 352,210 people in the Kumarkhali Upazila. Chapra

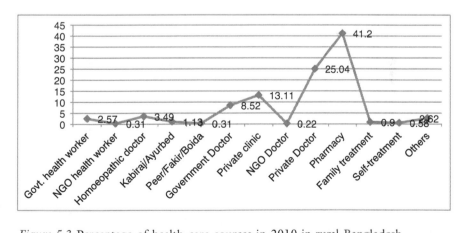

Figure 5.3 Percentage of health care sources in 2010 in rural Bangladesh
Source: Bangladesh Bureau of Statistics (2010)

residents can also access health services at two Chapra Union Sub-Centers (Ministry of Health and Family Welfare 2014). These public health services are not enough to ensure the health and well-being of marginalized people at Chapra. Abonti and Keramot informed me that one health officer with rudimentary training, operates one center and they are not able to provide treatment for serious illnesses like arsenic poisoning. The center is less than one bedroom apartment in size or about 2 square meters.

In addition to the health services provided by the above institutions, the central government has introduced the following seven health care safety-net programs: (1) Maternal Health Voucher Scheme, (2) National Nutrition Program, (3) Micro-nutrient Supplementation, (4) Revitalization of Community Health Care Initiative in Bangladesh, (5) National Sanitation Project, (6) Special Program for Irrigation and Water Irrigation and (7) Maternity Allowance Program for Poor Lactating Mothers (Figure 5.4). These programs were developed based on the Poverty Reduction Strategy Paper and Millennium Development Goals guidelines (International Monetary Fund 2013:143; Ministry of Planning 2013).

Local governments implement these programs for different target groups. The maternal health voucher scheme and maternity allowance programs seek to assist mothers who do not have the economic capability to maintain good health during pregnancy. In 2012–13, US$933,965,000 was spent nationally on this program. Nutrition and micro-nutrient programs (numbers 2 and 3 above) represent a significant effort to overcome malnutrition in poor children. However, 90 percent of focus group respondents report not receiving any safety-net program benefits.

Borkat, Fazlu, Kabir and Mofiz argued that, in this case, too, local elites control the central government's safety-net programs and exclude most marginalized people at Chapra from receiving benefits. Tanvir did not have any problem obtaining health services in the Kumarkhali Upazila Health Complex. Medical staff are always available for him and other local elites. But many of

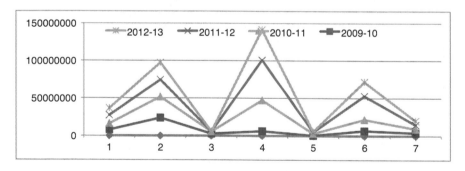

Figure 5.4 The central government's Health Care Programs from 2009–10 to 2012–13

Source: Ministry of Finance (2012)

my focus group respondents stated that doctors sell hospital medicines and use government medical technologies illegally for personal gain. They also do private consultancies even during hospital office hours. Consequently, marginalized people like Billal, Fazlu, Kabir and Mofiz receive inadequate or no care. Sometimes they are forced to buy medicine from local pharmacies, even when these medicines are supposed to be available for free to hospital patients.

Housing program management

Housing is an issue for many poor households in Chapra due to the limited availability of building materials like wood, which could formerly be obtained for free in the local environment, and due to the fact that, many people have lost their homes due to river bank erosion or displacement by GK canal infrastructure. Low wages and limited employment opportunities also make it difficult to buy the corrugated iron they generally use for roofs.

The central government has implemented six housing safety-net programs for poor people throughout Bangladesh: (1) Housing Support, (2) *Ashrayan* Project, (3) *Gucchagram*, (4) One Household One Farm, (5) Comprehensive Village Development and (6) Char Development and Settlement Project. As indicated in Figure 5.5, significant amounts of money are spent on these programs. Ministry of Disaster Management and Relief (MoDMR) provides housing grants to the Director General of Disaster Risk Reduction, which then allocates housing credits to the district councils throughout the country. The Deputy Commissioner of each district then distributes the housing credits among Upazila Councils and the Upazila Nirbahi Officer (UNO) is then directly responsible for distributing these credits to the eligible people.

The different housing safety-net programs focus on the different services. The ashrayan project provides assistance to the displaced people. This project builds houses in a specific area where local governments rehabilitate them. The central government increased this project budget from US$4,069,600 in 2009–10 to US$22,415,200 in 2012–13. The Gucchagram assists people who are

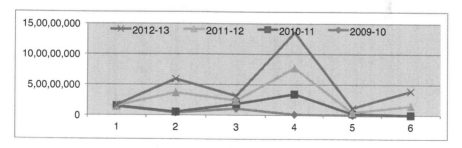

Figure 5.5 The governments' housing safety-net budgets (in US$)

Source: Ministry of Finance (2012)

considered to be climate victims. The budget for this program was US$10,646,000 in 2009–10 but was reduced to US$7,755,640 in 2012–13. The one household, one farm program provides eligible people with a small piece of land and money for housing materials. The budget for this program increased from US$996,229 in 2009–10 to US$58,528,500 in 2012–13. The government has also introduced a char development program to settle marginalized people on charlands but the Gorai River charlands at Chapra are not included in this budget. Although overall annual expenditures were several million dollars, these programs assist only a small number of all those who are in need. Also, as is the case with the other programs described above, local elites are able to capture many of these benefits for themselves, thus reducing the effectiveness of the programs still further.

Joardar reports having visited UC officials several times about his housing situation but received no help. He lost his ancestral home in 1960, when the government expropriated some of his land for construction of the Ganges-Kobodak Project. He built another house on the edge of his cropland that was destroyed by river bank erosion in 1987. He then applied to the GK authority for permission to erect a dwelling on their land and was allowed to build a small shack on a flood protection right-of-way, where he now lives along with his wife and children. Tanvir, meanwhile, is one of the major beneficiaries of local governments' housing programs. In addition to his house at Chapra, Tanvir has a house in Kushtia City and one in Dhaka that are used by his children. According to Joardar, Tanvir does not follow government procedures regarding housing safety-net benefits. He puts fake names on the beneficiary list and sells the housing benefits for personal profit. As a result of these illegal appropriations, Billal is also unable to receive housing assistance. He and his family currently reside within 6 meters of the Gorai River and face imminent displacement concerns and loss of housing due to flooding, embankment failure and river-bank erosion.

Borkat, Jobbar, Joardar and Billal also stated that local elites exploit displaced people by offering them informal housing opportunities on government lands, such as those belonging to the GK project, and then collect tolls from them or require them to work for no wage. Sometimes a member of an elite family will force a displaced person to commit criminal activities, such as smuggling or trafficking, in return for a place to live. Bakar, Fajal and Joardar stated that vulnerable people will also be coerced into political activities, including the intimidation or beating of supporters of other parties. After a few years, however, they are generally evicted from their informal housing and sometimes their houses will be burned down.

Conclusion

Local governments fail to perform proper roles in protecting the marginalized communities at Chapra due to top-down domination over the social protection programs. The central government promotes their own understanding of social protection programs and asks local government to implement them which fail to incorporate voices of the marginalized people. These programs fail to address locally embedded understanding of livelihood challenges which create further challenges for the marginalized people at Chapra. Moreover, the different resource management committees at district, *upazila* and *union* level are under direct control of the top-down government systems and power structures. The majority members of the committees are government staffs who are controlled by the officials of the UP and UC based on informal "blessings" of local MP. Some poor people are able to get these blessings in exchange for fulfilling some interests of the powerful. The major concerns behind this system is that it creates major gaps between social protection programs and community voices. For example, the fertilizer and seed management committee promotes chemical fertilizers and HYV crops that fail to recognize importance of ecological fertilizers and local seeds. Again, they want to overcome chemical fertilizer contamination from local water bodies based on local resource management. However, local community voices are excluded from the government programs due to top-down system that causes major concerns over basic rights: food, employment, education, health care and housing.

Based on this argument, in this chapter, I have documented the consistent failure of local government institutions to alleviate the chronic problems faced by the majority of rural households living in Chapra. These hardships begin with water shortages, ecosystem service failures and reduced crop productivity. They are made worse by agricultural development programs that are controlled by local elites working in concert with national political leaders and national and foreign corporations. The development programs benefit wealthy and intermediate farmers but reduce employment opportunities for the majority of households. These economic crises in turn lead to food shortages, health problems, low levels of educational achievement and inadequate housing for large numbers of people. Even though hundreds of millions of dollars are spent annually on safety-net and agricultural subsidy programs, the scale of the problem is well beyond the capacity of these programs to remedy and the little good they might do is undermined by corruption and manipulation by local elites. Government leaders, planners, aid agencies and international development banks represent the chronic problems faced by Chapra residents as the result of insufficient development but, in the next chapter, I propose an alternative theory, one that focuses on these problems as human rights abuses caused by too much development in the service of ecocracy.

Notes

1　Inheritance of land is usually through the male line. Females acquire rights to land only though marriage and upon marriage normally go to live on the land of their husband's family.
2　Bangladesh is divided into seven divisions and each division is further sub-divided into several districts. There are a total of 64 districts in the country.

References

Agrawal, A. 2003. Sustainable Governance of Common-Pool Resources: Context, Methods, and Politics. *Annual Review of Anthropology* 32: 243–262.

Agrawal, A. and E. Ostrom 2001. Collective Action, Property Rights, and Decentralization in Resource Use in India and Nepal. *Politics and Society* 29(4): 485–514.

Bangladesh Bureau of Statistics (BBS) 2005. *Key Information of Agriculture Sample Survey.* Dhaka: Government of Bangladesh.

—— 2010. *Area Population Household and Household Characteristics.* Dhaka: Government of Bangladesh.

Bardhan, P. 1993. Symposium on Management of Local Commons. *Journal of Economic Perspectives* 7(4): 87–92.

Bhuyan, O. U. June 15, 2013. *Study Finds Cadmium in Rice Samples.* Dhaka: New Age.

Ferguson, J. 1994. *The Anti-Politics Machine: Development, Depoliticization, and Bureaucratic Power in Lesotho.* Minneapolis: University of Minnesota Press.

Hayami, Y. 1981. Induced Innovations, Green Revolution, and Income Distribution: Comment. *Economic Development and Cultural Change* 30(1): 169–176.

Heyman, J. M. 2004. The Anthropology of Power-wielding Bureaucracies. *Human Organization* 63(4): 487–500.

Hossain, M. F. 2006. Arsenic Contamination in Bangladesh – An Overview. *Ecosystems and Environment* 113(1–4): 1–16.

International Monetary Fund (IMF) 2013. *Bangladesh: Poverty Reduction Strategy Paper (PRSP).* Washington: IMF.

Khan, A. M. 2013. The Genie Gave the Money. http://www.thedailystar.net/print_post/the-genie-gave-the-money-3009, accessed December 20, 2013.

Kushtia Zila Samity 2012. Education Institution Kushtia. http://kushtiazsdhaka.org/Educationalinstitution.php, accessed May 12, 2014.

Ministry of Agriculture (MoA) 2009a. *District and Sub-District Agricultural Rehabilitation Committee.* Dhaka: Government of Bangladesh.

—— 2009b. *Coordinated Policies for Fertilizer Dealers Appointment and Management,* trans. Dhaka: Government of Bangladesh.

—— 2011. *Coordinated Small Scale Irrigation Management.* Dhaka: Government of Bangladesh.

—— 2012. *Free Seed Distribution for Monsoon Seasonal Paddy Production.* Dhaka: Government of Bangladesh.

Ministry of Food and Disaster Management 2012. *Operation Manual 2012–13: Employment Generation Program for the Poorest.* Dhaka: Government of Bangladesh.

Ministry of Disaster Management and Relief (MoDMR) 2010. *VGF Food Crop Allocations for Ramadan Month,* trans. Dhaka: Government of Bangladesh.

Ministry of Finance (MoF) 2012. *Safety Net Programmes.* Dhaka: Government of Bangladesh.

—— 2013. *Safety Net Programmes.* Dhaka: Government of Bangladesh.

Ministry of Food (MoF) 2013. *National Food Policy Plan of Action and Country Investment Plan Monitoring Report 2013.* Dhaka: Government of Bangladesh.

Ministry of Food and Disaster Management 2010. *National Plan for Disaster Management 2010–2015.* Dhaka: Government of Bangladesh.

———— 2012a. *Operation Manual 2012–13: Employment Generation Program for the Poorest*. Dhaka: Government of Bangladesh.

———— 2012b. *Employment Generation Program for Hard-core Poor: 2010–2011 (2nd Phase)*. Dhaka: Government of Bangladesh.

Ministry of Health and Family Welfare (MoHFW) 2013. *Kushtia (Sadar) Upazila Health Office: Health Bulletin 2013*. Dhaka: Government of Bangladesh.

Ministry of Planning 2013. *Sixth Five Year Plan (FY2011-FY2015): Accelerating Growth and Reducing Poverty, Part-1 Strategic Directions and Policy Framework*. Dhaka: Government of Bangladesh.

New Age 2014. Upazila Polls: Most Candidates are Businessmen: Report. http://www.newagebd.com/detail.php?date=2014–02–18&nid=84203#.U69QjO_QfIU, accessed March 1, 2014.

Rahman, M. M., M. R. Rahman and M. Asaduzzaman 2010. Establishment of Dams and Embankments on Frontier River of North East Part of India: Impact on North-Western Region of Bangladesh. *Journal of Science Foundation* 8(1&2): 1–12.

Raihan, M. 2013. Political Party Supporters in Government Food Management Committee by Violating Rules, trans. http://nagoriknews.net/index.php/bangladeshi-newspaper/bangla-newspaper/ittefaq.html, accessed June 29, 2013.

Scott, J. C. 1998. *Seeing Like a State: How Certain Schemes to Improve the Human Condition Have Failed*. New Haven and London: Yale University Press.

6 Human rights at Chapra

In Chapra, access to a sufficient and predictable supply of water is the prerequisite for realizing other human rights to food, employment, health care, education and housing. The ability to realize these human rights is undermined by the effects of Farakka diversion constructed by India and the highly centralized, neoliberal approach to water management and agricultural development in Bangladesh. Political and technological interventions in the Ganges dependent area have resulted in the economic marginalization of the vast majority of households living in Chapra and in many other similar rural Ganges Dependent Area (GDA) communities. My goal in this chapter is to evaluate the degree of hardship experienced by marginalized households in relationship to international human rights agreements and international customary law in respect to water. If it can be demonstrated that these hardships do constitute human rights failures then the governments of India and Bangladesh, as the main agents of marginalization, can legitimately be held responsible for these failures.

I rely on several United Nations agreements and conventions in my analysis and on the Berlin Rules created by the International Law Association. Most important among the many applicable UN agreements are the 1948 Universal Declaration of Human Rights, the 1966 International Covenant on Economic, Social and Cultural Rights, a 2002 amendment to the 1966 Covenant referred to as General Comment 15, the 1997 United Nations Convention on the Law of the Non-Navigational Uses of International Watercourses and the 2010 General Assembly Resolution on the Right to Water and Sanitation. The 1948 UN Declaration of Human Rights recognizes basic necessities like food, employment, education and housing as human rights. The 1966 Covenant, which came into force in 1976 and was subsequently ratified by the governments of both Bangladesh and India, provides additional details about rights to education, health and an adequate standard of living. In 2002, General Comment 15, which deals explicitly with water and the relationship of water to other human rights, was added to the Covenant. The 1997 UN Convention on International Watercourses specifically addresses the responsibilities of states in respect to international rivers. The General Assembly Resolution of 2010 reaffirms the human right to water, citing all previous relevant agreements and covenants approved by the General Assembly. The Berlin Rules on

International Water Resources, which was finalized in 2004 by the International Law Commission, is the most comprehensive statement globally in respect to customary international water law. The 2004 Berlin Rules were developed as a revision of the 1966 Helsinki Rules in light of changes to international law including those brought about by 1997 United Nations Conventions on International Watercourses. Unlike the 1966 UN Covenant, neither the 1997 UN Convention nor the 2010 General Assembly Resolution have been ratified by either India or Bangladesh; however, international customary law does have legal standing in all countries and in international courts even when it is not formally recognized by a given country and no local enforcement mechanisms are in place.

Human rights violations in Chapra are inter-generational when families are displaced permanently from their ancestral homes with no chance of recovering them. Moreover, these violations are mutually reinforcing since lack of food, for instance, causes malnutrition and health problems which in turn further increase the causes of unemployment and a deepening cycle of food insecurity.

In this chapter, I describe in detail the nature and extent of the hardships being experienced by Chapra villagers in respect to the six human rights identified in Table 6.1 (water, employment, food, education, health and housing). I conclude with a consideration of how a better understanding of the relationship of river water to human rights might contribute to the restoration of human rights through community empowerment initiatives and improved policies for environmental governance. Table 6.1 demonstrates the percentage of households in my survey that I consider to fall below basic human rights standards in the six areas included in this study. These figures are based on responses to several survey questions about the status of these rights in Chapra. The survey was not designed primarily to facilitate the quantitative analysis of human rights violations but by summarizing response rates to several questions in each area of concern, it is possible to provide rough estimates of the percentage of households

Table 6.1 Human rights violations at Chapra

Human Rights	*Percentage of Households Falling Below Minimum Standards*
Sufficient water for basic needs	79
Adequate level of employment	77
Adequate level of food consumption	85
Reasonable access to formal education	79
Reasonable access to health care	66
Secure housing	73

Source: Author's research data

experiencing violations. The factors used to assess violations are provided in a series of tables below, one table for each human right category that is being assessed. Each factor within a table corresponds to one survey question. The response figures for each factor are averaged within each table in order to provide an estimate of the percentage of households experiencing human rights violations in each of the six human rights categories under study. Table 6.1 provides the summary results of these calculations.

Based on Table 6.1, I begin with a discussion of water as a human right and then examine, in turn, employment, food, education, health care and housing.

Water rights violations

My evidence, as summarized in Table 6.2, indicates that the water rights of the majority Chapra households, as defined by several international agreements and conventions, are being violated. The most important provisions clearly being violated are: (1) those pertaining to access by all citizens to "sufficient" and "affordable" water for basic needs, including agricultural subsistence; (2) the obligation of one state to not harm another by unilateral withdrawals of water from a shared river; (3) the special responsibilities of all states towards vulnerable communities; (4) the rights of affected people to participate in decision-making processes; and (5) the protection of "the ecological integrity" of a watershed. I discuss these factors in this order below, presenting quantitative information from my household survey when available and otherwise relying on qualitative information from focus group and case study informants.

Table 6.2 Water rights violations at Chapra

Factors Preventing Access to an Adequate Supply of Water	*Percentage of respondents reporting this problem*
Insufficient local water supply due to reduced flow in the Ganges River	100
Water Stagnation problems	62
Gorai River channel change problems	90
River bank erosion problems	84
Embankment failure problems in 2011	63
GK Project water supply problems	68
GK Project's source water point affected by the Ganges water reduction	84
Water rights violations (average)	*79*

Source: Author's research data

Water access for basic needs

Article 17 of the Berlin Rules (International Law Commission 2004:17) provides one of the clearest legal statements on "the right of access to water." Article 17.1 states that "every individual has a right of access to sufficient, safe, acceptable, physically accessible, and affordable water to meet that individual's vital human needs." Article 17.2 states that right of access must be available on a "non-discriminatory" basis. Article 17.3.a specifies the need for the state to not interfere in an individual's exercise of their right to water. Similar water rights definitions are included in articles 11 and 12 of the United Nations General Comment 15 (2002).

Data presented in chapters three, four and five demonstrate that a majority of poor households in Chapra experience chronic shortages of the water required to grow subsistence crops; the water used for drinking, cooking and bathing is regularly contaminated by agricultural chemicals and pathogens associated with standing water; water access is highly discriminatory; the states of both India and Bangladesh have interfered with people's abilities to access the water they need for basic human needs. Ninety-five percent of my survey respondents informed me that proper water access failures are responsible for agricultural failures. Their primary source of water was formerly the Gorai River and the failure of this river forced them to depend on GK Project or tube-wells. However, 84 percent of my survey respondents believe that the Ganges Basin flow reduction due to the Farakka diversion causes the GK Project water supply to be insufficient. Furthermore, 78 percent of survey respondents fail to get sufficient support from GK Project staff to resolve their water problems. Neither did they get any cooperation from the local water management association. Local elites control the water management association in such a way that 74 percent of marginalized households report not receiving minimum information about their water practices. When they fail to get affordable water access from rivers, streams or GK canals they are forced to use purchase water from tube wells. Despite this alternative mechanism, 63 percent of my survey respondents report experiencing agricultural production damages and losses due to water supply problems. Their suffering increases further with environmental vulnerabilities like water stagnation. Among my survey respondents, 62 percent report water stagnation problems in their neighborhoods, which cause multiple problems like health and crop production concerns. These problems are escalated with the river bank erosion, which is encountered by 76 percent of my survey respondents. Among them, 44 percent encountered this erosion three to four times in their lives. Sixty-one percent of my survey respondents do not have access to safe drinking water sources such as uncontaminated tube wells or hand-dug wells known as kuya. Instead, they use sources like stagnant water that is often contaminated with agricultural run-off. Ninety-one percent of my survey respondents report no longer practicing traditional sports and festivals due to the Gorai River water failures. Some of these celebrations, like boat races, occur less often because of reduced water flow, while the others, like pitha (sweetmeat) festivals, occur less often because of crop production failures and a lack of cash to buy the requisite ingredients.

State obligations in regard to shared watercourses

As documented in chapter two, the flow of Ganges River water to Bangladesh during the dry season is now less than one-third of its volume prior to the construction of the Farakka Barrage and about two-thirds of its volume during the wet season. This is a major cause, though not the sole cause, of water shortages in Chapra. This represents a clear infringement by India of article 7.1 of the UN Convention (United Nations 1997) and articles 12.1 and 16 of the Berlin Rules (International Law Association 2004:20, 22). Article 7.1 states that "watercourse States shall, in utilizing an international watercourse in their territories, take all appropriate measures to prevent the causing of significant harm to other watercourse States." The government in India violates these articles not only through the operation of the Farakka Barrage but also through the operation of a whole series of hydro-infrastructures throughout the Ganges Basin in India (Khan 1996). Article 12.1 of the Berlin Rules specifies that, "basin States shall in their respective territories manage the waters of an international drainage basin in an equitable and reasonable manner having due regard for the obligation not to cause significant harm to other basin States." Article 16 of the Berlin Rules states that, "basin states, in managing the waters of an international drainage basin, shall refrain from and prevent acts or omissions within their territory that cause significant harm to another basin State having due regard for the right of each basin State to make equitable and reasonable use of the waters." Due to violation of these articles, the flow of water in the Gorai River, a distributary of the Ganges, is reduced well below the one-third level of the Ganges itself due to sedimentation that is particularly intense around the mouth of the river. Chapra villagers during the dry season do not receive the minimum amount of river water they need for agricultural production. Only 4 percent of my household survey respondents report being able to access Gorai River water for agricultural production. However, 100 percent of local people used this river and other people distant from the river, report using local ponds, wetlands and lakes before the Farakka diversion.

Since the government of Bangladesh has not been able to prevent India from withdrawing significant amount of water from the Ganges River water, due to the nature of hydro-politics in the region (Bhattarai 2009; Brichieri-Colombi and Bradnock 2003; Elhance 1999; Hossen 2012; Iyer 2008; Nakayama 1997; Paisley 2002), they have failed in their obligations to "respect," "protect" and "fulfil" the water rights of their citizens as outlined by the Berlin Rules (International Law Association 2004). Additionally, the government of Bangladesh is responsible for worsening water shortages through its own top-down water management programs like the Ganges-Kobodak and Gorai River Restoration Projects.

Vulnerable communities and the right to participate in decision-making processes

Data presented in the preceding chapters also makes it abundantly clear that the process of marginalization of smallholder farmers and farm laborers has occurred as a result of their exclusion from decision-making processes and the ability of local elites to control local water management institutions and decision making. Class relations have a long history in this region and some degree of marginalization and poverty was certainly in place before the 1970s. Marginalized households, however, did not previously exist in the numbers they do today and the scale of previous hydrological interventions was not such that people could be displaced or impoverished in such large numbers. With the encouragement of international donor organizations, the government has introduced a number of safety-net programs to address the needs of vulnerable communities but have done nothing to reverse the underlying cause of their vulnerability. This clearly contravenes Article 20 of the Berlin Rules which reads:

> States shall take all appropriate steps to protect the rights, interests, and special needs of communities and of indigenous peoples or other particularly vulnerable groups likely to be affected by the management of waters, even while developing the waters for the benefit of the entire State or group of States
>
> (International Law Association 2004:26).

Chapra residents are also entitled to compensation based on article 21 of Berlin Rules that specifies "duty to compensate persons or communities displaced by water projects or programs" (International Law Association 2004:27). This compensation has international and national dimensions. Local people in the Ganges Dependent Area should receive the compensation from the Government of India due to the harms created by the Farakka diversion. For example, India paid this compensation to Pakistan (Parua 2010:252). In addition to this international aspect, the government in Bangladesh creates much harm to local communities with water and agricultural development programs. My data documents the fact that many households have lost land due to project interventions and have been entirely displaced on some occasions with little or no compensation. Forty-three percent of my survey respondents report losing lands to the GK Project construction and 83 percent report losing more than 1 acre of land. The government in Bangladesh provided compensation for GK Project expropriations; however, this compensation was controlled by local elites and often did not find its way to marginalized people. The government did not provide any compensation to the victims of the Gorai River bank erosion. This erosion caused the loss of more than 3 acres of land for 46 percent of my survey respondents.

Ecological integrity

River water rights violations at Chapra create ecological resource rights violations. As noted throughout this book, the natural resources formerly available to Chapra residents as common property resources are vital to their sustainability. Several articles in the Berlin Rules and the 1997 UN Convention use the concepts of ecological integrity and ecological flows to define the nature of the customary water rights in this domain. Article 3.6 of the Berlin Rules defines ecological integrity as "the natural condition of waters and other resources sufficient to assure the biological, chemical, and physical integrity of the aquatic environment" (International Law Association 2004:9). Article 3.1 of the Berlin Rules describes the aquatic environment as "all surface and groundwater, the lands and subsurface geological formations connected to those waters, and the atmosphere related to those waters and lands" (International Law Association 2004:9). Article 20 of the 1997 UN Convention provides guidelines for "the protection and preservation of ecosystems" and states that "watercourse States shall, individually and, where appropriate, jointly, protect and preserve the ecosystems of international watercourses" (United Nations 1997).

Article 15 of the Berlin Rules emphasizes that water rights include protection of "water necessary to assure ecosystem services or otherwise to maintain ecological integrity or to minimize environmental harms" (International Law Association 2004:22). Reduced river flows at Chapra reduce local ecosystem services like cropland siltation, wild fisheries, wild vegetables and vegetation. All of my survey respondents unanimously argued that local water bodies like Chapaigachi oxbow lake, Lahineepara, Shindha, and Shaota canals, and Lahineepara and Chapra Fakirapara wetlands are facing survival challenges due to the Ganges-Gorai River ecosystem failures. The demise of these water bodies violates article 22 of the Berlin Rules that reads: "States shall take all appropriate measures to protect the ecological integrity necessary to sustain ecosystems dependent on particular waters" (International Law Association 2004:28). Furthermore, article 24 of the Berlin Rules specifies that "States shall take all appropriate measures to ensure flows adequate to protect the ecological integrity shall take all appropriate measures to ensure flows adequate to protect the ecological integrity of the waters of a drainage basin, including estuarine waters" (International Law Association 2004:29).

The current management system also contributes to the creation of "harmful conditions" such as more extreme and unpredictable flooding, prolonged drought and river bank erosion (Table 6.3). Article 27 of the 1997 UN Convention describes the obligation of states for the "prevention and mitigation of harmful conditions" that include flood, erosion, sedimentation, salinity and drought (United Nations 1997). Article 32 of the Berlin Rules describes the need to protect citizens against "extreme conditions . . . that pose a significant risk to human life or health, of harm to property, or of environmental harm" (International Law Association 2004:33). Article 21(2) of the UN Convention notes; "watercourse States shall take steps to harmonize their policies in this

Table 6.3 Harmful ecological effects as reported by case study participants

Harmful Conditions	Smallholders and Landless Laborers		Wealthy and Intermediate Landowners	
	Billal	Joardar	Tofajjal	Tanvir
Loss of land/crop	X (no)	√ (yes)	√	√
Water access failure	√	√	X	X
Soil depletion & nutrient loss	X	√	√	X
Embankment failure	√	√	√	√
River bank erosion	√	√	√	√
Flood damage & loss	√	√	√	√
Drought damage and loss	X	√	√	√
Water stagnation problems	√	√	X	X

Source: Author's research data

connection" so that no "harm to human health or safety" will occur (United Nations 1997). Article 28 of the 1997 UN Convention further describes the responsibility of states to "immediately take all practicable measures necessitated by the circumstances to prevent, mitigate and eliminate harmful effects of the emergency" (United Nations 1997). Due to the Ganges Basin management failures, 62 percent of my survey respondents report facing water stagnation problems that create disease pathogens, mosquitoes and loss of production. The unregulated use of chemical fertilizers promoted by the state also violates laws that require states to minimize environmental harm by controlling pollution.

My analysis of water as a human right also incorporates principles regarding the importance of water rights to many other human rights. General Comment 15 emphasizes that priority in water allocation should be for "personal and domestic uses" but Article 1.6 also states that: "water is required for a range of different purposes, besides personal and domestic uses, to realize many of the Covenant rights. For instance, water is necessary to produce food (right to adequate food) and ensure environmental hygiene (right to health). Water is essential for securing livelihoods (right to gain a living by work) and enjoying certain cultural practices (right to take part in cultural life)" (United Nations 2002). Article 2.11 further states that, "the adequacy of water should not be interpreted narrowly, by mere reference to volumetric quantities and technologies. Water should be treated as a social and cultural good, and not primarily as an economic good" (United Nations 2002).

In the sections that follow, I assess the extent to which human right failures in respect to water create a cascading effect on several other human rights and on their cultural practices in general. Kusum and Jobbar's question is very important here to understand marginalized people's human rights concerns: *"Paani jodi na thaake tahole amra ki kore bajbo, abad kibabe hobe, kaaj kothai pabo?"* (How can we survive if we fail to have water, how can we cultivate crops and get work?)

The right to employment

Article 23.1 of the UN Human Rights Declaration states the following: "everyone has the right to work, to free choice of employment, to just and favorable conditions of work and to protection against unemployment" (United Nations 1948). Article 1.6 of General Comment 15, as noted above, links water rights to livelihood rights (United Nations 2002). Article 13.2.b of the Berlin Rules explicitly links water rights to "economic needs" which include the right to employment (International Law Association 2004:21). Ninety percent of my household survey respondents informed me that obtaining sufficient local employment is dependent on success in traditional agricultural production, whether on one's own land or through employment on another's land. Successful agricultural production generates employment opportunities for farmers, day laborers and household workers through the many activities described in previous chapters, for instance, planting and harvesting crops, care of livestock, boat driving, bullock cart driving, fishing, paddy husking, irrigation and the making of traditional fertilizers. This production promotes the local economy which generates employment opportunities for other traditional occupational groups like blacksmithing. Fifty-six percent of my survey respondents report facing loss of employment in 2010 extreme enough to categorize as a violation of their human right to employment (Table 6.4). Forty-seven percent of survey respondents report being unemployed for more than two months in 2010 due to ecological vulnerabilities like flooding and drought. The major causes of these employment rights violations are loss of traditional agricultural practices as a consequence of induced technological change. This problem is then compounded with the exclusion of marginalized people from government employment programs due to local elite domination.

Only the households I am classifying as marginalized face employment rights violations at Chapra, since the better-off households have been able to protect or expand their assets over recent decades. Tanvir, for instance, reports earning an average of US$2,151 a month throughout 2010,[1] whereas landless day laborers like Billal earn, on average, a maximum of US$67 (Bangladesh Bureau of Statistics 2010a). Billal earned this amount based on his own and other household members' earning on a daily average wage of US$1.88. Compared with Billal, Joardar, a smallholder, has better economic opportunities and can earn US$100 a month based on a combination of self-employment and wage labor employment. Tanvir reports earning US$5,000 in 2011 from a single fish project, and

Table 6.4 Employment rights violations at Chapra

Factors Preventing Adequate Access to Employment	Percentage of Respondents Reporting this Problem
Employment opportunities lost due to river water reduction	87
Loss of traditional employment opportunities	51
Employment opportunities lost due to new technologies	91
Secondary (off-farm) employment now essential to survival	82
Significant periods of unemployment in 2010	56
Exclusion from the government's Food For Work program	93
Employment rights violations (average)	*77*

Source: Author's research data

he has two more fisheries projects in addition to commercial agricultural operations. He also operates other businesses selling chemical fertilizers and HYV seeds, and selling or renting tractors, deep tube-wells and mini-transport vans.

My fieldwork data demonstrates that there is a huge gap between the income and expenditures of an average marginalized family. The Bangladesh Bureau of Statistics (BBS) reports that one marginalized person, like Billal, can earn a maximum of US$67 per month based on current employment opportunities but that he needs at least US$230 per month for his nine-member family to meet food, education, health care and housing needs (Figure 6.1). These calculations, made by the government's own statistical bureau data, demonstrate that a violation of human rights is occurring in this setting, as defined by article 23.3 of the UN Declaration of Human Rights: "Everyone who works has the right to just and favorable remuneration ensuring for himself and his family an existence worthy of human dignity, and supplemented, if necessary, by other means of social protection" (United Nations 1948). Ninety one percent of my survey respondents indicated that new agricultural technologies have reduced employment opportunities at Chapra. This pattern is especially damaging for elderly people (those over 60) who, according to the Bangladesh Bureau of Statistics (2012), constitute 9 percent of the population of the Kumarkhali Upazila. The elderly cannot develop the new skills needed to operate agricultural and irrigation machinery. Disabled people, 1.3 percent of the local population, are also severely disadvantaged by the move to more technology-based employment opportunities.

The story of Kamal, a focus group respondent, is illustrative of the hardships being faced by Chapra residents. He explained to me that he had been a smallholder farmer but had lost his land due to GK Project construction and river

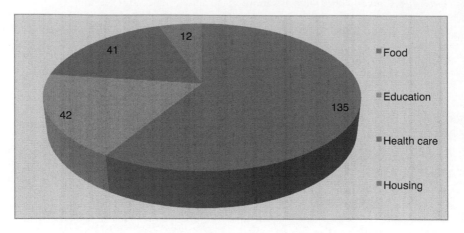

Figure 6.1 Minimum expenditures (US$[1]) of a marginalized household at Chapra

Source: Bangladesh Bureau of Statistics (2010a)

1 US$1= BDT 69.23, October 12, 2010

bank erosion and now made his living as a day laborer. Currently, he does not get enough work in Chapra, despite searching out every possible job opportunity, for example, as a rickshaw puller, mason, fisherman or poultry-farm worker. He migrates to a city during the rainy season for four months, while local employment opportunities are fewest. Abonti, Kamal's wife, works as a domestic servant in the house of Ibrahim, a wealthy Chapra landowner, and works part-time along with her daughters in jute and pepper processing activities owned by Mokles, another wealthy farmer.

Women like Abonti and Billal's wife, Parvin, do not raise questions about wages, working hours and job tasks with employers or the government out of fear of losing their jobs. There are 298,366 women like Abonti and Parvin in Kushtia District, who make their livings mainly as low paying workers, earning less than $6 a month (Bangladesh Bureau of Statistics 2012). Parvin starts work early in the morning and continues until late at night. She works on a no-work-no-pay basis and does not receive weekend or overtime benefits. Parvin stated that many people fear being killed by their employers if they demand reasonable wages. For example, an Akij Bidi factory owner in Kushtia is alleged to have killed two workers and injured others when the factory workers were demonstrating for higher wages on July 15, 2012 (Afroz 2012). The most vulnerable employment groups are women, disabled, elderly people, widowed and children laborers. Parvin informed me that widow maids occasionally need to "satisfy" employers for job security or normal life in a locality.

Employment rights violations include child labor exploitation. Many children lost a father or mother who was the main income earning member of a household. Or in some cases the household head is encountering chronic illness or disability. The children are then sent to search for employment opportunities. Many of them work as transport helpers and others work at poultry farms and are forced to sell diseased chickens to local hotels or neighborhoods. Chickens that die from various illnesses are routinely sold illegally in this manner. Tanvir employed Billal's son, Lablu, to sell pest-damaged vegetables at local markets, thus making him a potential target for verbal or physical abuse if people discover the problem. Lablu is only paid if he is able to sell the vegetables and avoid recrimination.

Unemployment problems lead marginalized people to work under conditions best understood as slave labor. Kamal cannot afford to take a sick leave or a weekend break. No employers take even a minimum economic responsibility for their employees when they encounter sickness or lose a body part, like a hand. No women employees get maternity leave with financial benefits. Pope Francis himself termed Bangladeshi garment workers as "slave labor" because of their low wages and poor working conditions (*Guardian* 2013). The worse case scenario is that the garment workers' wages and benefits are better than the agricultural workers at Chapra. The garment workers are paid extra for weekend work and are allowed a leave of absence for vacations that are absent in rural Chapra. Many of agricultural workers need to work at charlands despite major environmental and physical challenges like cyclones and drought, food deficiency, sickness and transportation. Charlands do not have proper road or water transportation systems causing major challenges to reach the work place with agricultural production materials like tractors. Many workers died at work and their families have failed to get compensation. For example, two people died and another fell sick in Kushtia when they were cleaning a rooftop water tank of a residential building on June 11, 2011 (*Daily Sun* 2011). Many marginalized groups, for example, the disabled, widowed, elderly people and female workers encounter the worst employment rights violations. Focus group respondent Suman informed me that the practice of employing bonded laborers is increasing in Chapra, a clear violation of article 4 of the UN Human Rights Declaration (United Nations 1948:3): "no one shall be held in slavery or servitude." People are forced into bonded labor after failing to repay a debt of some kind, for instance, money borrowed to construct a house.

Eighty-three percent of my household survey respondents report taking loans of one kind or another in order to ensure their survival. Tofajjal has a small business at Chapra where he sells daily necessities like rice, wheat, biscuits and fuel. Joardar borrowed daily necessities like rice from Tofajjal in 2011, leaving three silver plates as security. Tofajjal and Joardar made a verbal commitment that Tofajjal would return the plates after Joardar paid back the amount. According to Tofajjal, Joardar failed to keep his commitment because of unemployment. The crop losses due to flooding in 2011 and irregular GK water supplies in 2012 made Joardar incapable of paying the loan. Therefore, Joardar failed to get back the plates and lost US$74. Currently, 81 percent of my survey respondents like Joardar and Billal report being loan defaulters due to unemployment.

Unemployment generates desperation and a search for alternative survival strategies at Chapra. Billal informed me that Begum, a local widow, sold her child in 2012 for US$184 to an outsider to pay her NGO micro-credit loans and reduce her past household loan burdens. Two people at Chapra sold their kidneys for similar purposes. Many others sell blood to buy food when they face prolonged periods of starvation. They are desperate to get any job, legal or illegal. Many work as paid political activists on behalf of local political leaders. Political leaders will pay the equivalent of $3 per day for people to engage in organizing or attending political rallies on their support. Focus group respondent Fajal informed me that Tanvir demanded Fajal help him gain control of newly developed Gorai River charlands in 2012 through violent means. Men hired by Tanvir and those hired by a wealthy local rival, Faisal, were involved in a brutal conflict on January 20, 2014, to gain control over these lands. This conflict is reported to have resulted in 1 person dead and 11 injured (*UNBconnect* 2014). Marginalized people at Chapra are forced into these actions simply to secure employment opportunities. In Dhaka, poor children and women are hired by some political leaders to make their political programs a success in exchange for money. The Dhaka Tribune (2013) reports that "they [political leaders] paid a teenager to carry a petrol bomb from one place to another for BDT 100 (US$1.20) and BDT 300 (US$3.60) for a blast and BDT 500 (US$6.20) for an explosion in a public place or in a vehicle." Sometimes, law enforcement agencies execute such people extra-judicially because of the illegal nature of the activities. For example, police killed two alleged "kidnappers" during an encounter at Mirpur Upazila in Kushtia on November 23, 2011 (*New Age* 2011). These desperate income-generating activities are not sustainable and create multiple socioeconomic problems. They provide ample evidence, however, of the fact that the right to employment, as defined by the United Nations, is not being realized by the majority of Chapra residents.

The right to food

Article 25(1) of the UN Human Rights Declaration inform us "everyone has the right to a standard of living adequate for the health and well-being of himself and of his family" and that sufficient food is one essential component of an adequate standard of living (United Nations 1948). As stated previously, General Comment 15 also affirms "the right to adequate food" in the context of its declaration that water is a human right partly because "water is necessary to produce food" (United Nations 2002). Food rights are also considered "vital human needs" as identified in Article 14.2 of the Berlin Rules (International Law Association 2004) and Article 10.2 of General Comment 15 (United Nations 2002). The 1997 UN Convention does not explicitly mention water but as noted in the commentary of Article 18 of the Berlin Rules, the UN General Assembly approved a statement of understanding in 1997 that declared, "special attention is to be paid to providing sufficient water to sustain human life, including both drinking water and water required for the production of food in order to prevent starvation" (United Nations 2002:4). Agricultural crops and essential ecological resources like water, natural fertilizers, wild vegetables and fisheries are diminishing day-by-day in Chapra and food has become a scarce resource for many. During a focus group discussion, Laily stated that "*maache vaate bangali*" (abundant rice crops and river fisheries are a historical tradition of the Bengali). Focus group respondents Rahim, Mofiz and Bimal informed a common past statement about local fish availabilities: "*motso maribo khaibo shuke*" (available river fish as common ecological resource is helpful for nutrition and happiness). However, this access is no longer available and fish is currently a major commodity. Eighty-one percent of my survey respondents report food shortages that I classify as food rights violations based on the human rights documents cited above (Table 6.5).

Sixty percent of my survey respondents report having to reduce the number of daily family meals they eat. Kamal reported that he eats two meals or less rather than the regular three meals a day and that these two meals are not even full thus he does not achieve minimum nutrition criteria. This was commonly reported by other informants from marginalized people as well.

Tanvir and Tofajjal, on the other hand, have surplus food supplies as a result of their land-ownership, agricultural technologies and commercial operations. On the other hand, Joardar, because of limited landholdings and unemployment, cannot earn the minimum monthly amount of US$83 needed to feed his six-member household. Billal's condition is worse than Joardar. He needs US$135 a month for his nine-member household, although his only income source is traditional employment like agricultural laborer, river fishing or boat driving. Table 6.5 delineates the minimum food cost per household based on Bangladesh Bureau of Statistics (BBS) (2010a). BSS data estimates of nutritional

Table 6.5 Food rights violations at Chapra

Factors Determining Adequate Access to Food	Percentage of Respondents Reporting this Problem
Food security is not achieved every year	86
Food security is not realized most of the time	81
Debt level compromises food access	81
Fish consumption levels reduced by comparison to the past	91
Have not received government school food program benefits	84
Food rights not secured (average)	*85*

Source: Author's research data

Table 6.6 Monthly average per capita food expenditures at Chapra in 2010

Major Food Items	Per Capita Intake (in kg)	Household Intake (5.5 people)	Cost in BDT
Cereals (rice, wheat and others)	14.57	80.14@27	2164
Potato	2.15	11.83@12	142
Vegetables	5.10	28.05@15	421
Pulses (masoor, khasari and others)	0.40	2.2@48	106
Milk/milk products	0.96	5.28@45	238
Meat, poultry and eggs	0.62	3.41@200	682
Fish	1.32	7.26@80	581
Spices (including onion, chilies, etc.)	1.94	10.67@23	246
Fruits	1.28	7.04@65	458
Sugar/gur	0.22	1.21@40	49
Other (tea, bread, betel leaf, etc.)	1.00		90
Total Food Cost in Bangladesh Taka			5,177
Total Food Cost in US Dollars			**$75**

Source: Adapted from Bangladesh Bureau of Statistics (2010a)

requirements and the cost of food items as recorded in the Lahineepara market in Kushtia in November 2011.

In addition to the food items, Billal needs to buy other household items like water, cooking fuel, clothing and utensils. Moreover, a religious or social celebration like Eid or Christmas has a special food demand for family members. For one Eid celebration, a marginalized household like Billal needs US$25. He cannot make special purchases for Eid without adjusting with food budget reduction for a longer period. Billal and Joardar are representative of 92 percent of the people in Kushtia who do not have the opportunity to enjoy social gatherings like religious celebrations, marriage ceremonies and community festivals; these deprivations generate social problems like divorces and suicides. For example, a father is reported to have killed his two children and attempted to commit suicide on August 19, 2012 at Bheramara in Kushtia when he was unable to buy new dresses for his children's Eid celebration (*Reflection news* 2012).

Kamal pointed out that the different categories of marginalized people encounter different severities of food rights violations, depending on their socioeconomic capabilities (Table 6.7). Most disabled people are unemployed and many of them eat less than a full meal once a day. Most of them become disabled due to employment hazards, road accidents and health care failures. Widows (8.6 percent of the sub-district population) and the elderly (8.7 percent of the population) face malnutrition and even starvation in disproportionate numbers to the remainder of the marginalized population in Kumarkhali Upazila. Kafil reported that many elderly people cannot walk

Table 6.7 Categories of people most at risk of food violations in Kumarkhali Sub-district

Categories of People Most at Risk	Total	As Percentage of Sub-district Population		
		Total	Male	Female
Disabled	4,270	1.3	–	–
Widows	11,186	8.6	0.6	8.0
Divorced/separated	909	0.7	0.1	0.6
Elderly people (60+ years)	28,576	8.7	–	–
Homeless (beggars)	117	0.1[1]	–	–
Domestic servant	51,912	44.3	0.3	44.0
Totals	97,355	63.9	–	–

Source: Bangladesh Bureau of Statistics (2012)

1 The number of homeless people is significantly under-reported according to information gathered through other sources including personal observation.

properly and beg in the street for survival. Many of them collect *vater-maar* (broth produced from cooked rice) from neighbors. He saw destitute children searching for food in garbage bins and competing with dogs for a piece of bread. Many of these now destitute people had abundant food supplies in their past lives that they describe as a time of *"gola bhora dhan, goal bhora goru, pukur bhora match"* (surplus paddy at storage places, cattle at homesteads and fish in ponds). Ninety-one percent of my survey respondents report insufficient supplies of fish protein due to the reduction of wild fish populations in local water bodies (Table 6.5).

Focus group respondent Raju reports undertaking many desperate acts to secure sufficient food, acts that can be described as violations of Article 1 of the UN Declaration of Human Rights: "all human beings are born free and equal in dignity and rights" (United Nations 1948). Raju compromises the quality of the food he eats out of desperation for securing at least minimum food access. He and his family eat cadmium-affected hybrid rice because of its lower price. As noted previously, the overuse of triple superphosphate fertilizer in Kushtia District causes the contamination of rice by an average of .099 milligrams of cadmium per kilogram (Bhuyan 2013). Raju and his whole family do not eat meat or milk for several weeks at a time because they are unable to maintain their own domestic animals and cannot normally afford to buy these food items. He tries to obtain meat no one else wants, such as a piece of stomach from a slaughtered cow, to get taste of meat. Tanvir does not eat the stomach and throws it into the garbage after processing the slaughtered animal. Raju asks Tanvir in advance for a portion of the stomach. In fact, he competes with other marginalized people, like Billal and Fajala, to gain a portion of it. Tanvir even exploits this competition by asking Raju to process the animal without any wage in exchange for the stomach. Tanvir provides this chance only for his supporters. Other focus group respondents, like Fajal, report collecting dead chickens from garbage bins or chicken farms for household meat consumption. Sometimes, they buy diseased beef from local sources for less than 50 percent of the regular price. They also collect pest-affected or rotten vegetables, fish and fruits for lower prices. These consumption practices are against their self-dignity and constitute a food rights violation as defined by the United Nations.

The information provided by Billal and Parvin demonstrates the strongly gendered nature of food rights violations at Chapra. Parvin, Billal's wife, informed one of my female research assistants that Billal has full household responsibilities and therefore he should be healthy, physically and mentally. I, along with Billal, heard this conversation from a few feet away and Billal agreed with this statement. Based on this understanding, Billal and his children eat at first and Parvin eats afterwards if there is any food leftover. As a result, she suffers from chronic malnutrition. Billal eats the best piece of fish and his children share the left over portion. Parvin will often eat only the left over bones of the fish after serving her husband and children. Sometimes, she serves all the available food to Billal and her children and does not even keep bones for herself. She hides food pots so that Billal cannot see that they are empty. When Billal asks about her food intake, she replies with a smile, so that he does not feel bad for her starvation. Again, Billal asks his children about starvation and their responses are that they are not hungry. Currently, Billal's children, like other marginalized children, are encountering malnutrition and are chronically underweight. This is a common issue throughout Bangladesh where medical testing indicates that 48 percent of infants are underweight (Matin et al. 2009).

Food rights violations at Chapra are especially hard to bear by marginalized people, given their intimate knowledge of the circumstances of wealthy families in their own community. While Billal and Rahim face chronic malnutrition and even starvation, Tanvir regularly wastes significant amounts of food. Tanvir with his full family and friends can afford to eat out at fast food restaurants spending more than US$100 for a single meal. Tanvir's family is able to feed two dogs, one cat and other domestic animals, like chickens, almost entirely on food waste. Parvin, who works in Tanvir's house, manages to get a portion of this waste for her children and husband, Billal. Food deprivation sometimes leads to extreme tensions and social problems within marginalized households when husband and wife confront each other due to starvation and malnutrition and such confrontations will be witnessed by relatives and neighbors. Women are especially vulnerable in these situations and, in an extreme situation, a wife will leave her husband even at the risk of her own life. On November 3, 2012, for example, a woman was reported to have been found murdered, floating in the GK canal at Chapra, following a domestic dispute (Akram 2012).

Education rights violations

Unemployment and food rights violations lead in turn to education rights violations for marginalized people. Article 26 of the UN Declaration of Human Rights states that "everyone has the right to education. Education shall be free, at least in the elementary and fundamental stages" (United Nations 1948). Article 19 of the Berlin Rules also includes guidelines in respect to education rights, since citizens need a certain level of education to participate in water management processes "States need to undertake to educate their people to enable them to fulfill their rights and duties as citizens and as stakeholders in the management of waters" (International Law Association 2004:26) However, my data indicates that 55 percent of my survey respondents are experiencing education rights violations (Table 6.8). Ninety-two percent of survey respondents reported that their children's access to education depends on successful agricultural production. This production provides the foundation for yearly education costs. In Kushtia, 33 percent of all boys and 26 percent of all girls drop out of school between primary (I-V) and secondary (VI-X) grade levels (Bangladesh Bureau of Statistics 2011). Many parents do not even enroll their children in the primary grades in order to reduce their household economic burdens.

As noted in the previous chapter, children from marginalized families do not usually succeed in school for two main reasons: (1) they do not attend on a regular basis because their families cannot afford school costs and because they are an essential source of labor for their families; and, (2) marginalized households cannot provide children with adequate study conditions at home due to lack of lighting and crowded living conditions. The two factors inevitably lead to high drop-out rates. In Kushtia, the average literacy rate is 46

Table 6.8 Education rights violations at Chapra

Factors Preventing Adequate Access to Education	Percentage of Respondents Reporting This Problem
Access to education is dependent on agricultural production	92
Unable to set aside a yearly education budget	84
Unable to afford all educational expenses as they come due	82
Unable to remain in school until at least grade 10	55
No benefit from NGO contribution to education costs	84
Failure to achieve basic access to education (average)	*79*

Source: Author's research data

percent; among them 48 percent are male and 45 percent are female (Bangladesh Bureau of Statistics 2012:1). There are no school fees in Bangladesh but each family nevertheless faces an average school expense of US$6 per month for each child according to the Bangladesh Bureau of Statistics (2010a:131). These costs include school uniforms, books and other supplies, and transportation or accommodation costs for children in secondary schools since there is no high school in Chapra. Billal has seven children and, therefore, requires a minimum of US$42 per month to send all his children to school. Marginalized households like Billal clearly do not have the financial means to secure even a minimum standard of education for their children. When I asked him about access to formal education for his children, he expressed a deep concern with a comment "*bhaat paina, shikhya dibo kibabe?*" (How can I afford education for children when I do not have enough food for the whole family?). Focus group participant Kusum stated that, currently, her family's spending priorities are for irrigation water supplies, chemical fertilizers, HYV seeds and pesticides for agricultural production. These new economic burdens have thus made children's education expenses a lower priority than they once were for many marginalized families.

This is in sharp contrast to the better-off Chapra households, many of whom send their children to private schools that follow Western educational curricula and use the English language as the medium of instruction. They also send their children to developed countries like Canada for higher education. Tanvir emphasized the idea that success in higher education promotes social mobility. This access to education promotes access to power structures. Masud, Tanvir's oldest son, is a joint secretary, a high-level bureaucratic position, of the Bangladesh Government. Masud's wife has an MSc degree and works for a local bank. Tanvir's youngest son, Firoj, migrated to Australia and Firoj's wife also holds an MSc degree and works for the Australian government. Tanvir's educated children develop social networks through marriage, school and professional networks. Based on these networks, Tanvir has a connection to the national level of government in Bangladesh and through his connections, he is able to control local water and agricultural management systems.

In contrast, marginalized people are excluded from local power structures due to their low level of education. They are stigmatized as less knowledgeable, backward or low caste people and fail to maintain social prestige. Socially marginalized households from occupational castes like sweepers or blacksmiths are routinely discriminated against by school administration staff. Physically disabled or widowed people are also among those least able to afford education for their children. Furthermore, there are no proper schools, teachers or education materials for the children who disabled. Consequently, these youth remain illiterate and uneducated and face major survival challenges; they are officially landless and live on charlands or at other illegal sites, many work as bonded laborers and die at work places, and females are forced into early marriage.

During previous generations there has been a greater tendency for well-off families to assist poor students with their costs for food and accommodation, a system locally known as *jaigidar*. Likewise, school teachers formerly extended more care to weaker students, even visiting their homes when they were absent from school. These practices no longer exist; currently, teachers are forced to supplement their incomes with commercial activities like private tutoring. Sometimes, they exploit school schedules for personal interests, teaching illegally as part-time employees of private schools. Neoliberal approaches to education by the Bangladesh Government are leading to an increase in private schools and fewer public schools. As noted in the previous chapter, education safety-net programs have been created to provide more help for marginalized students but they are manipulated by the local elite and teachers themselves so as to exclude many eligible recipients. Borkat informed me that, due to flood damage in 1988, the local primary school held classes in an open field under a shade tree. Borkat further informed me that this local primary school was closed for renovations for six months, due to flood damages in 2007. Many schools in the region have closed forever, for example, the MN High School in Kushtia, which was destroyed by Gorai River bank erosion. The flood-displaced people took shelter in the school building during the flood period and, due to this fact and disruptions to transportation and communication systems, teachers and students were unable to carry out regular educational activities. Educational standards in this setting are thus very sensitive to extreme weather events of many kinds, flooding as well as the more pervasive influence of chronic water shortages and associated loss of ecological resources and crop productivity. Given the anthropogenic nature of these problems, it is reasonable to argue that the inadequate levels of education at Chapra constitute human rights violations for which the state itself is responsible.

The right to health

Employment, food and education rights violations caused by water mismanagement and ecological resource failures also lead to health rights violations at Chapra. Article of 25.1 of the United Nations Declaration of Human Rights states the following (United Nations 2002):

> Everyone has the right to a standard of living adequate for the health and well-being of himself and of his family, including food, clothing, housing and medical care and necessary social services, and the right to security in the event of unemployment, sickness, disability, widowhood, old age or other lack of livelihood in circumstances beyond his control.

Article 12.1a of General Comment 15 affirms that everyone has the right to sufficient water for "drinking, personal sanitation, washing of clothes, food preparation, personal and household hygiene" and that health is one criterion for evaluating the adequacy of water supply (United Nations 2002). Health

Table 6.9 Health rights violations at Chapra

Factors Preventing Access to Adequate Level of Health Care	*Percentage of Respondents Reporting This Problem*
Health is not maintained well enough to avoid crop production failures	74
Unable to set aside a yearly budget for health care services	83
Do not have a safe supply of drinking water	67
Do not have a sanitary latrine	29
Local sanitary facilities not sufficient to avoid health problems	37
No access to subsidized NGO health services	91
Health care is inadequate to family's basic needs	81
Health care rights not protected (average)	*66*

Source: Author's research data

rights, which require adequate amounts of clean drinking water and sanitation services, are considered "vital human needs" as defined in article 14(2) of Berlin Rules and article 10(2) of the 1997 UN Convention on International Watercourses. On the basis of these definitions, I calculate that 81 percent of my survey respondents at Chapra are experiencing health rights violations (Table 6.9).

Many marginalized people at Chapra are experiencing chronic sicknesses like fever, dysentery, gastric ulcers and tuberculosis (Bangladesh Bureau of Statistics 2010a). My own research also confirms they are subject to malnutrition and stunting, diarrhea, arsenic and cadmium poisoning and skin diseases. The immediate causes of the vast majority of these illnesses are unclean drinking water, chemical pollution, no sanitation facilities, food shortages and consumption of unhealthy food. Almost all of these causes arise in turn from the mismanagement of water resources and from water-dependent agricultural systems. Ninety-six percent of my survey respondents reported that their health is mainly dependent on successful agricultural production. When they are able to save this production from flooding, drought or pest attack, they are able to make some profits which provide the major foundation for health care expenditures.

Focus group respondent Kabir was one of many who reported health rights violations that arose due to food rights violations. As a result of insufficient land and income, Kabir consumes the meat from sick, dying or dead animals which causes health concerns and even has resulted in sudden death for some. He cannot afford fresh fish so buys cheaper fish that died from chemical contamination. Tarun and Bakar report health problems caused by the consumption of cheaper vegetables that are nearly rotten or contaminated by chemicals. Because of the amount of chemically contaminated food they consume, according to Keramot, their longevity is less than that of previous generations. These problems are in addition to the many problems described in the right to food section above, such as stunting, which are associated with chronic malnutrition.

Twenty-nine percent of my survey respondents also report not having access to sanitary latrines at their homes. Many of them use non-sanitary facilities and open spaces to dispose of waste that, in turn, contributes to the lack of safe drinking, bathing and cleaning water sources and create problems like diarrhea and skin diseases. Sixty-seven percent of my survey respondents report experiencing drinking water problems during periods of flooding or drought conditions. Kabir reported using the same water as a toilet and a source of cleaning, bathing and drinking water during the flood of 1988.

Many of the "harmful conditions" that result in health right violations in this setting originate from the top-down water and agricultural management system. As Article 8 of General Comment 15 makes clear, the state has a responsibility to protect its people against "harmful conditions" arising from both sanitation and chemical pollution (United Nations 2002):

> Environmental hygiene, as an aspect of the right to health under article 12(2), paragraph 2 (b), of the Covenant [on Economic, Social and Cultural Rights], encompasses taking steps on a non-discriminatory basis to prevent threats to health from unsafe and toxic water conditions.

Government introduced chemical fertilizers and pesticides, in support of HYV crop production are also significant sources of health problems. In Kushtia, the total fertilizer usages of urea, TSP, DAP and MP had increased 934 percent, 127 percent and 343 percent respectively from 2008–09 to 2011–12 (Ministry of Agriculture 2012). As noted previously, cadmium contamination of rice is now a major health issue but local crops and plants are also encountering new types of fungi, insects and toxic pollutants that generate health problems. In 1999, 180 grams of pesticide was used for every hectare of cropland in Bangladesh (Ministry of Agriculture 2002). These fertilizers and pesticides contaminate local water bodies based on their spread by rainfall and flooding.

Focus group respondent Rahim pointed out that, health rights violations produce multiple social problems. Seventy-two percent of my survey respondents reported missing work days and encountering crop production failures due to sickness. The health rights violation of one person in a household will have a multiplier effect since other household members may then fail to get enough food because of another's illness. Rahim suffered from jaundice for more than three years. The doctor told him to take rest and eat nutritious food to recover from this sickness. However, he did not have the opportunity to take rest because his whole family depends on his earnings for basic needs like food. Further, his wife has been hospitalized due to pregnancy-related health problems, which was an additional burden for them. Due to their economic problems, his wife is at extreme risk from maternal death. Consequently, he suffers from depression and fails to secure minimum health and well-being rights. These multidimensional effects of health rights violations reduce his ability to afford other basic necessities.

Government health programs do very little to alleviate the health problems of marginalized households. As noted previously, landless laborers like Billal can earn US$67 monthly on average but need US$177 for food and education expenditures alone. This leaves a gap of US$110 without even considering health care costs that, on average, will require an additional US$41 monthly per household on the basis of data provided by the Bangladesh Bureau of Statistics (2010a) (Table 6.10).

According to the Bangladesh Multiple Indicator Cluster Survey 2009, 41 out of every 1,000 children die in Kushtia District during the first year of life and an additional 52 children die before the end of their fifth year (Bangladesh Bureau of Statistics 2009:17). In 2011, Kusum suffered an attack of tuberculosis that lasted 12 days. She has been sick on and off with tuberculosis for more than two decades and is unable to recover completely due to inadequate health care and treatment. Towards the end of my research period, her health condition deteriorated further due to the combined failures of healthcare services and nutrition.

As described in the preceding chapter, the doctor-patient ratio is declining day-by-day in Kushtia and, in 2011, there were 352,210 people depending for treatment on the 12 doctors and 50 beds of the Upazila Health Complex in Kumarkhali. The doctor-patient ratio, in other words, is 29,351, and there are 7,044 people for every hospital bed (Ministry of Health and Family Welfare 2012). In addition to this poor ratio, marginalized people are routinely discriminated against by medial staff and safety-net programs are controlled by elite groups with very few benefits going to those most in need. Medical infrastructure itself is also not well maintained. The union health care center, in

Table 6.10 Estimated per capita expenditures of a marginalized person for one month

Items	Cost in Bangladesh Taka
Consultation fee	156.53
Hospital/clinic charges	394.43
Cost of medicine	248.81
Cost of medical tests	226.08
Transportation costs	90.66
Other	254.32
Total	**1,400.00 (US $41)**

Source: Bangladesh Bureau of Statistics (2010a:298)

addition to community clinics and the Kumarkhali Upazila Health Complex, are at risk from Gorai bank erosion. The floods of 1988, 1999, 2004 and 2007 inundated local health care service centers. Buildings, machinery and hospital beds have all been damaged by the flooding. Floods and embankment failures also disconnect local people from health care service centers. Kabir's father died from diarrhea because he was unable to reach the Upazila Health Complex due to flooding in 1988.

The right to housing

As noted in the previous section, article of 25.1 of the United Nations Declaration of Human Rights affirms that housing is a human right necessary to the achievement of an adequate level of health and well-being. Article 11.1 of General Comment 15 identifies the right to water as a prerequisite for securing the right to housing (United Nations 2002). The right to housing is included among the "vital human needs" identified in article 14.2 of the Berlin Rules (International Law Association 2004) and in article 10.2 of the UN Convention on International Watercourses (United Nations 1997). However, 86 percent of my survey respondents report what I consider to be housing rights violations (Table 6.11). Currently, thirty-three percent of smallholder and landless respondents do not have their own houses and live in informal housing, sharing place or open land they do not own. Respondents also made clear their opinions on the close relationship of housing to agricultural production and the agro-ecological system (Table 6.11).

Currently, many marginalized people have no choice but to build their houses along the Gorai River bank area which is at major risk of damage due to erosion, embankment failure and flooding. To avoid flood inundation in 2007, Keramot reports having raised his sleeping platform, locally called

Table 6.11 Housing rights violations at Chapra

Right to Housing	*Percentage of Respondents Reporting This Problem*
Traditional housing materials cannot still be gathered for free	77
Do not hold ownership rights to family home	33
Insufficient yearly budget for renovations	76
Not secure against displacement by government projects	86
Has not received government housing program support	93
Housing rights not protected (average)	*73*

Source: Author's research data

maccha, with bamboo and rope. Many others followed a similar strategy and covered this maacha with fishing net to avoid snake bites. Keramot saved few emergency resources like cooking pots, domestic chickens and cooking fuel with this raised platform as well. The living space was limited, so female members slept in this raised place and male members of the household slept elsewhere during this time. Keramot passed nights on his boat and his son lived at a neighbor's house. After the flood period, he was not able to repair his house and thus had to put up with leaks from his roof as well as damage to bed, clothing and stored food. He continues to be at high risk, since his house is 25 meters from the Gorai River.

Seventy-seven percent of my survey respondents report having changed their housing patterns due to the Ganges flow reduction. Earlier, they were successful in getting house-building materials like wood, bamboo, straw or rope from local ecological resources that were almost freely available and environmentally sustainable. Borkat argued that his traditionally constructed house was cooler during the summer season and warmer during the winter season. However, community people no longer have these ecosystem friendly housing patterns. As noted in the previous chapter, the central government assisted in the distribution of new housing technologies using materials like cement, corrugated iron and iron rods but most marginalized people cannot afford these materials. According to the Bangladesh Bureau of Statistics and my own research, the average marginalized household now needs to spend a minimum of US$12 a month to secure access to safe housing (Table 6.12). This cost only includes maintenance and rental costs, excluding new construction costs, which are much higher. For example, materials to build a rural house at Chapra the size of an one bedroom apartment, with a mud floor and corrugated iron walls and roof, costs US$700. This cost is beyond the economic capability of a marginalized household. The central government's housing safety-net programs have also failed to provide meaningful assistance to poor rural households for the reasons outlined in the previous chapter.

Currently, the housing gap between rich and poor is increasing (Table 6.13). Marginalized households like Sabbir's can only build a *kutcha* house that has a roof made of bamboo and straw and walls and floor made from a local clay

Table 6.12 Average housing expenditures of a marginalized rural household in 2010

Social and Well-being Items	Expenditures (BDT)
House rental or maintenance costs	687
Household expenses (utensils, table, chair)	100
Total	**787 or US$12**[1]

Source: Bangladesh Bureau of Statistics (2010a) (house rental) and Focus Group data (household expenses)

1 US$1= BDT 69.23, October 2010

Table 6.13 Housing types in Kumarkhali Sub-District
(percentage)

Types of House	1991	2011
Kutcha	39	76
Semipucca	59	21
Pucca	2	4

Source: Bangladesh Bureau of Statistics (1991, 2011)

mud that hardens when it dries. Intermediate landowners like Tofajjal can afford to build a *semipucca* house that has roofs and walls made of metal sheets and wood and a floor made of brick or clay. Rich people like Tanvir can build a *pucca* house which has walls, roof and floor built with cement and brick. They also build the floor of their houses with cement and brick. The number of people in Kumarkhali Sub-district with kutcha housing, the poorest form of housing, rose from 39 percent in 1991 to 76 percent in 2011, while the number of households with intermediate semipucca housing dropped from 59 percent to 29 percent (Bangladesh Bureau of Statistics 1991, 2011). The number of highest quality pucca houses doubled during this period from 2 percent to 4 percent. The BBS data thus supports my findings that there is an overall trend towards poorer quality housing except among a small number of elite households.

Many marginalized people at Chapra move to cities because of displacements from their ancestral home and seek work as garment workers or rickshaw pullers. In Bangladesh, about 35 million people live in city slums, a number projected to increase rapidly to 68 million by 2015 (Azad 2013). In Dhaka, 40 percent of total people lives in slums or on the streets and 20 percent of them fail to find a place to live in even in the slums and live on the streets along with their wife and children (Islam 2012). In addition to city slums, more and more people now live in rural slums, on charland and along river banks. These displaced people face the most extreme levels of housing rights violations. Many of those growing up in this situation are displaced from their family and lose all contacts. In Bangladesh, 400,000 children are orphans who live in streets, among them, 20,000 are girl sex workers (Jaijaidin 2014).

Well-off people like Tanvir, however, can afford to build a house with excellent facilities, safe from ecological vulnerabilities and connected to roads and highway. He and his family live a modern lifestyle in a home that features tiles, sculptures, paintings and professionally designed architectural features. Masud, Tanvir's oldest son, designed this house with an architect friend and brought materials from Dhaka. Firoj, another son, brought some other housing materials from Australia. This new house protects rich people like Tanvir from ecological vulnerabilities and provides comfort and security. He has an additional two houses in Kushtia and Dhaka.

Many rich people sell or lease rural assets like cropland, fruit gardens and fisheries projects and move to a city and establish new businesses. The rich urban migrants appoint caretakers to manage agricultural and fisheries production and invest their rural profits in the city. For this purpose, Tanvir sends his children to Kushtia and Dhaka cities. His children's identities are thus changing from farmers' children to government servants or businessmen. They are part of a new urban middle class and major figures in the national power structure. In the rural areas, elite families of this type maintain control over local power structures and attempt to win positions such as Member of Parliament (MP) or Minister. Changes to local housing patterns are thus associated with the upward social mobility of the rich and housing rights violations for the poor.

Focus group respondent Harun reported experiencing a series of displacements and emphasized how each displacement added new socioeconomic burdens. When forced to sell household assets and croplands, he failed to get reasonable prices. He needed to sell these assets to local elites for *joler dame* (an incredibly lower price) because of his extreme situation. Since he would have had to pay more than double the price he received in order to replace the same assets at a new place, he ended up without adequate housing.

In many cases, as noted above, people in severe distress move from rural to urban settings. In 2011, 8 percent of people migrated from rural to urban areas in Bangladesh (Bangladesh Bureau of Statistics 2011). When they go, they often leave elderly and disabled family members behind and thus become estranged from their extended family network as well. They are likely to face new housing rights violations living in informal slum housing and law enforcement agencies often drive them out from one place to another. Corrupt political leaders harass women in these informal housing situations and victims are afraid to seek legal remedies for fear of losing this shelter. This and other forms of exploitation produce hatred against political leaders and loss of respect for government and the law, together with a loss of dignity and hope for a decent life. This dignity is described as social customs and traditions defined as human rights by the United Nations.

Conclusion

The right to water of the vast majority of Chapra farming households, as defined by United Nations Conventions and international customary law, is being systematically violated and that this violation in turn has led to multiple human rights violations in the areas of employment, food, education, health care and housing. Due to this water right failure, Chapra villagers fail to get ecological resources like cropland siltation, water-borne wild vegetables, and local seeds. The basin communities at Chapra fail to get regular *borsha* flow that creates failures of ecological fertilizers like siltation, algae, earth-worm and water hyacinth. Therefore, the failure of this siltation creates challenges for agricultural production and employment opportunities. The central government in Bangladesh executed technology-based water and agriculture management to overcome this effect which create the differential outcomes between Tanvir, Tofajjal, Joardar and Billal. The central government established the GK project, HYV crop and chemical fertilizer to ensure agriculture production. The rich farmers, Tofajjal and Tanvir, promote commercial agricultural production and agri-business based on these technologies. On the other hand, the marginalized farmers, Joardar and Billal, encounter further survival challenges. The decrease of agricultural production and employment creates major challenges for spending of US$160 monthly on their food and well-being items. They can earn maximum US$67 that describes a major gap of US$93 to maintain a minimum food and well-being expenditure. In this context, they do not have scopes for spending US$42 and US$41 on education and health services respectively that create further human rights violations. Under the circumstances, the marginalized groups of people like physically disable, widow, elderly and chronically sick confront the worst human rights failures of food, employment, education, health care and housing. Many of these victim people displaced from their ancestral home that generated new level of human rights violations. They live in informal places like slum or embankment and are excluded from government services like water or sanitation. Therefore, the housing rights violations, together with employment, food, education, health and water rights violations are a major cause of the social movements that organize to counter the negative outcomes of government and elite control of local ecological resources. In the next chapter, I will describe different aspects of these local movements as they occur at Chapra based on my ethnographic fieldwork data and observations.

Note

1 Tanvir's real income would be much more than this since it would include many informal and even illegal activities as well as family business operations registered in the names of other family members.

References

Akram, S. M. 2012. The Body of a Woman Floating in GK Canal Kushtia, trans. http://dailyandolonerbazar.com/04/11/2012/news/details/nid=68709, accessed November 5, 2012.

Azad, T. M. A. 2013. City Ultra-Poor Livelihood. Dhaka: Daily Ittefaq. http://www.ittefaq.com.bd/index.php?ref=MjBfMTJfMDNfMTNfMV81XzFfOTAy NjU=, accessed December 3, 3013.

Bangladesh Bureau of Statistics (BBS) 1991. *Bangladesh Population Census*. Dhaka: Government of Bangladesh.

—— 2009. *Monitoring the Situation of Children and Women: Multiple Indicator Cluster Survey (MICS) 2009*. Dhaka: Government of Bangladesh.

—— 2010a. *Area Population Household and Household Characteristics*. Dhaka: Government of Bangladesh.

────── 2010b. *Household Income and Expenditure Survey (HIES)*. Dhaka: Government of Bangladesh.

────── 2011. *Population and Housing Census 2011*. Dhaka: Government of Bangladesh.

────── 2012. *Population and Housing Census 2011*. Dhaka: Government of Bangladesh.

Bhattarai, D. P. 2009. An Analysis of Transboundary Water Resources: A Case Study of River Brahmaputra. *Journal of the Institute of Engineering* 7(1): 1–7.

Bhuyan, O. U. June 15, 2013. *Study Finds Cadmium in Rice Samples*. Dhaka: New Age.

Brichieri-Colombi, S. and R. W. Bradnock 2003. Geopolitics, Water and Development in South Asia: Cooperative Development in the Ganges-Brahmaputra Delta. *Royal Geographic Society with IBG* 169(1): 43–64.

Daily Sun 2011. Two Killed While Cleaning Septic Tank in Kushtia. http://www.daily-sun.com/details_yes_21–06–2011_Two-killed-while-cleaning-septic-tank-in-Kushtia_257_1_10_1_19.html, accessed July 3, 2013.

Dhaka Tribune 2013. Hartal a Profitable Business for Some Poor Children, Women Used as Proxy Activists. http://www.dhakatribune.com/crime/2013/dec/06/hartal-profitable-business-some-poor-children-women-used-proxy-activists, accessed June 27, 2014.

Elhance, A. P. 1999. *Hydro-politics in the Third World: Conflict and Cooperation in International River Basins*. Washington: United States Institutes of Peace Press.

Guardian 2013. Bangladesh Factory Collapse: Pope Condemns 'Slave Labor' Conditions. http://www.theguardian.com/world/2013/may/01/bangladesh-factory-pope-slave-labour, accessed May 7, 2013.

Hossen, M. A. 2012. Bilateral Hydro-hegemony in the Ganges-Brahmaputra Basin. *Oriental Geographer* 3(53): 1–18.

International Law Association 2004. Berlin Conference (2004) Water Resources Law. http://www.internationalwaterlaw.org/documents/intldocs/ILA_Berlin_Rules-2004.pdf, accessed May 23, 2013.

Islam, S. 2012. Forty Percent People of Dhaka Megacity Lives in Slums. http://www.ittefaq.com.bd/index.php?ref=MjBfMTJfMTVfMTJfMV8xXzFfMzc1NQ==, accessed June 23, 2014.

Iyer, R. R. 2008. *India's Water Relations with Her Neighbors*. Washington: The Simon Center.

Jaijaidin 2014. Floated 20 Thousand Girl Children Are Sex Workers, trans. http://www.jjdin.com/?view=details&archiev=yes&arch_date=13–2–2014&type=single&pub_no=746&cat_id=1&menu_id=13&news_type_id=1&index=5, accessed February 15, 2014.

Matin, I., M. Parveen, N. C. Das, N. Mascie-Taylor and S. Raihan 2009. *Implications for Human Development-Impacts of Food Price Volatility on Nutrition and Schooling*. Dhaka: BIDS Policy Brief.

Ministry of Agriculture (MoA) 2002. *Pesticide Usage in Bangladesh*. Dhaka: Government of Bangladesh.

────── 2012. *Free Seed Distribution for Monsoon Seasonal Paddy Production*. Dhaka: Government of Bangladesh.

Ministry of Health and Family Welfare 2012. *Health Bulletin 2012; Kumarkhali Upazila Health Complex*. Dhaka: Government of Bangladesh.

Nakayama, M. 1997. Success and Failures of International Organizations in Dealing with International Waters. *Water Resources Development* 13(3): 367–382.

New Age 2011. One Killed, Ballot Boxes, and Paper Snatched. http://newagebd.com/newspaper1/archive_details.php?date=2011–06–04&nid=21290, accessed July 1, 2012.

Paisley, R. 2002. Adversaries into Partners: International Water Law and the Equitable Sharing of Downstream Benefits. *Melbourne Journal of International Law* 3(2): 1–21.

Parua, P. K. 2010. *The Ganga: Water Use in the Indian Subcontinent*. Dordrecht: Springer.

Reflection New 2012. Father Kills Minors for Demanding Eid Dress. http://reflectionnews.com/father-kills-minors-for-demanding-eid-dress/, accessed December 9, 2012.

UNBconnect 2014. Man Killed in Kushtia Clash. http://unbconnect.com/clash-killing-14/#&panel1–9, accessed January 21, 2014.

United Nations 1997. Convention on the Law of the Non-navigational Uses of International Watercourses. http://www.un.org/law/cod/watere.htm, accessed December 12, 2013.

——— 2002. General Comment No. 15 (2002): The Right to Water. http://www2.ohchr.org/english/issues/water/docs/CESCR_GC_15.pdf, accessed June 23, 2010.

7 Political resistance, conflict and social movements

In this chapter I discuss the social protest, movements and everyday forms of resistance that have emerged in Kushtia District in response to the hardships described in previous chapters. I also discuss the forms of resistance that have emerged in other local settings and at the national level in Bangladesh. To conceptualize the pattern of resistance that is occurring, I especially rely on Scott's theory concerning "weapons of the weak" and "everyday forms of peasant resistance" (1985). Patterns of resistance in Kushtia do not always achieve permanent or structured forms given the level and nature of the crises and the immediate objectives of the protests. The most vulnerable people occasionally take extreme actions against the people they hold responsible for their economic hardship, while less vulnerable people may use more traditional mechanisms or organize under the banner of a local or national social, political or community organization. There is a long history of public protest against the Farakka Barrage, for instance, and one of the most notable protests, the "long march" of 1976, was organized by a national level opposition party leader. The targets of public protests vary a great deal but the level of resistance is so extreme and spatially and temporally widespread that statecraft in Bangladesh faces an uncertain future if the political elite continue to dismiss and misrepresent the nature of the resistance.

Escobar's (1992) concept of "pluralistic grassroots movements" is also very relevant to the pattern of resistance in Bangladesh as is the concept of "subaltern social movements" as used by Kapoor (2011). In Chapra and Kushtia District specifically, however, local forms of resistance tend to be diffuse, spontaneous and "off-stage" except when they are organized by local chapters of national political parties and NGOs. Many of local movements have arisen largely in response to the effects of the Farakka Barrage, GK project and GRRP which have created livelihood challenges and human right violations described in the previous chapters. Furthermore, their exclusion from social service benefits creates frustration and hatred against the ruling elites responsible for the misdeeds. Local movements seek to overcome this exclusion and negative effects of water development projects. In the long run, their goal is to restore ecosystems, promote basin-wide water governance and to emphasize bottom-up water and agricultural development.

I begin my discussion of resistance with an overview of the current political climate in Bangladesh. Protestors are routinely dismissed by the government as either followers of an opposition party or as terrorists. Extra-judicial killings of protestors and protest leaders are all too common. I divide my discussion of resistance itself into those forms of resistance that focus mainly on resource issues and those that focus mainly on political processes. Whereas the former seek remedies for environmental degradation and loss of resources, the latter promotes democratization, political inclusion and bottom-up approaches to development.

Conceptualization of local movements

The in-depth case respondents Billal and Joardar informed me that they had glorious past of accessing ecological resources like river water, fisheries, vegetable, domestic animals, local seeds, natural fertilizer, and water-borne wild vegetables. However technocentric water development in India and Bangladesh reduces these resources and increase ecological vulnerabilities like flooding, drought, river bank erosion and salinity intrusion. Because of these negative effects, many Chapra residents are destitute, displaced from the local area, and have nothing more to lose from livelihoods and encounter extreme level of poverty. Their displacements from ancestral homes spread local movements from local to international levels. Wherever they move, their livelihood sufferings pile up due to continuing failures of human rights: food, work, education, health care and housing. For example, according to Billal, two victims of *Rana* Plaza garment collapse that occurred in 2013 hailed from Kushtia and this collapse killed more than 1,200 garment workers (Rashid 2013), many of whom are ecologically displaced people. Like these Rana Plaza victims, Tarun added, many rickshaw pullers in Dhaka are displaced from the Gorai River banks in Kushtia. Their sufferings from ecological vulnerabilities and displacements reminded them the causes like the effects of the Farakka diversion, GK and GRRP. Furthermore their informal residences in slums and other settlements cause exploitation of local *maastan,* or musclemen, and law-enforcing agencies. The *maastan* collect tolls from them and sometimes forces them to become traffickers and smugglers. If they refuse to get involved in this criminal activities, they encounter risk of eviction. Their life-long sufferings from the negative effects of the water development projects make them desperate people to develop strategies as major means of survivals.

Scott's approach is relevant to the situation occurring at Chapra very well when he describes the struggles of Malaysian peasants in the 1980s as struggles over "land, work, income, and power in the midst of the massive changes brought about by an agricultural revolution" (1985:49). Scott also writes that, "every subordinate group creates, out of its ordeal, a 'hidden transcript' that represents a critique of power spoken behind the back of the dominant" (Scott 1990:xii). The marginalized people at Chapra have indeed created a hidden transcript and give expression to it with the same "constant, grinding conflict over work, food, autonomy, ritual" that Scott describes as "everyday forms of resistance" (1985:5). Local people practice these everyday forms of resistance at Chapra due to the criminalized government system and the risk of being killed by law enforcement agencies and government supporters. As in the Malaysian case described by Scott, most of this resistance has "no name, no organization, no leadership, and certainly no Leninist conspiracy behind it" (1985:53). The culture and organization this encourages at the bottom is reactive and disorganized resistance that is Scott's main point. So, following up his earlier book published in 1979 about peasant "moral economy" and resistance in Vietnam in the 1960s/70s, he argues that the peasants lack proactive organization and are unable and moderately unlikely to mobilize proactively to advance their interests.

However, the resistance of people at Chapra in Kushtia District does not always follow the pattern emphasized by Scott. Smallholder farmers, agricultural laborers, fishermen and boatmen are also frontline protestors and demonstrators, participants in hartals and the petitioners of district, upazila and union level governments. Some people support political organizations that use violent methods to advance their goals. Local day laborers, women and children, on occasion, participate in public protests against their Member of Parliament (MP), the bureaucrats, local government members and rich farmers who exclude them from subsidies, safety-net programs and wild fisheries. They occasionally boycott elections and protest against political leaders who they know are involved in budget mismanagement, vote riggings, nepotism and criminal practices. Many will attempt to remain anonymous and unidentified during these protests but the "hidden transcript" is not entirely hidden. It is important to state outright that these reactive actions are typical of powerless populations. Since they lack the proactive means of carrying out demands for change, they resort to individualized and small group violence and obstruction. E. J. Hobsbawm's (1959) classic *Primitive Rebel* which explains the culture and organization quite well for medieval and early modern Europe has a lot of parallels here in terms of cultural outlook and lack of organizational capacity.

Moreover, many social movements exist in Bangladesh at local as well as national levels and some of those movements fit well with the concept of "subaltern social movements" as defined by Kapoor (2011). An excellent example in India is the Chipko movement, which promoted community control over forestry (Gadgil and Guha 1992; Rangan 2004:370). The concept of the subaltern applies somewhat differently in Bangladesh, however, where India has been the "colonizing" power since Bangladeshi independence. More useful is Escobar's concept of "localized, pluralistic grassroots movements" (Escobar 1992:27) and Turner's concept of "parapolitical" social formations (Turner 1997). Pluralistic movements have many goals, attract different groups of people with diverse objectives and distrust conventional political parties and 'establishment' organizations. Parapolitical social action, according to Turner (1997:10) is "action above and beyond the political system of the state towards universal social, ethical and cultural values." Two examples in Bangladesh of these types of movements are described below in more detail, the Beel Dhakatia and Beel Kapalia environmental movements that have protested against water development. In many cases, rural Bangladeshis ally themselves with organizations such as the UBINIG (Unnayan Bikalper Nitinirdharoni Gobeshona or Policy Research for Development Alternatives), which promotes bottom-up approaches and tries to mobilize local voices on a range of issues (Mazhar 1997). This is a large national level organization, however, and many senior staff are well connected politically. In some cases, organizations of this type end up constraining or even subverting local resistance movements as a consequence of their political affiliations and local people are therefore wary of such organizations, working with them on some occasions but avoiding them on others. This strategic relationship can be conceptualized with Migdal (1996) and Scott's (1979) approaches of local movements. Their main thesis is that the "moral economy of risk control" included engagement in traditional patron-client ties, so when patrons gradually withdrew from the local village, there was no reason to not rebel against the "disruptive patron." It does not necessarily lead to a more proactive conception of community but it is a factor in the process and leads to rebellion.

Protests over water and resource management are only one component, however, of the protest culture that is now so dominant in Bangladesh. This chapter concludes by describing the broader form of protest that seeks to reform the very foundations of the political system in Bangladesh and that addresses additional forms of human rights abuses other than those associated with the control of environmental resources. In this context, it is important to understand the government system that originated from the political climate in Bangladesh.

The political climate in Bangladesh

The current political culture in Bangladesh is not suitable for protecting marginalized people; they want to change this culture. The two major political parties of Bangladesh Nationalist Party (BNP) and Bangladesh Awami League (AL) are divided on national agendas based on their patrimonial and dynastic legacies. The BNP chief, Begum Khaleda Zia, is the widow of former President Ziaur Rahman and the AL chief, Sheikh Hasina, is a daughter of the founding President of Bangladesh Bangabandhu Sheikh Mujibur Rahman. In 2012, for example, the opposition BNP wants national election under caretaker government due to vote rigging concerns although the ruling AL government wanted this election under their supervision. This BNP refused to accept AL's caretaker system demand when they were in power in 1991–1995. Based on this party line, the different professional groups and associations develop power circles to fulfill their personal interests and fail to recognize voices of the majority people. In this context, clientelism is the organizing principle behind the leading parties and their dynastic families that run the party system as if it were a spoils system to be privately appropriated. Most of the corruption and mismanagement I discuss at various points that operates as a great source of grievance for the local people and the small and landless farmers.

Based on this power circle, most of the political parties want to stay in government power by any means as long as possible. For this purpose the government exploits bureaucratic organizations and fail to ensure equal rights and responsibilities for every group of people. Sometimes, criminals get impunity from punishments in exchange of supporting the ruling government. This political support develops new forms of political culture that is suitable for ruling elites. This culture can be described in the context of both parties as patron-client i.e., based on personalized and totalistic loyalties – which promotes a zero-sum conflict mentality throughout the system.

As consequences of this political culture, violent public protests have become the norm in Bangladesh over the past several years and are currently one of the most significant problems in the country. These conflicts take many different forms: *hartal* and *dharmaghat*. The term *hartal* is used to refer to the massive public demonstrations that shutdown transportation systems, schools, offices and shops and bring life in many parts so the country to a grinding halt. National level opposition parties organized 481 days of hartal against the government from 1991 to 2001 (Mithu 2011). Hartals were also very common during the period of the fieldwork in 2011–12 and violent in the year leading up to the last national election in January 2014. The term *dharmaghat* refers to strikes or other actions that involve a withdrawal of services by occupational groups such as garment workers. Street demonstrations organized by political parties and occupational groups are also very common. Dharmaghats and demonstrations are both much more frequent than hartals but are usually more localized and cause less disruption. The government usually attempts to convince the public that these protests are acts of terrorism by Muslim fundamentalists or international conspiracies. Law enforcement agencies are called on to stop the protestors and supporters of the party in power organize counter demonstrations and sometimes attack the protestors. Protestors are often killed and the violence can escalate to claim lives on both sides. For example, fifteen people were reported killed in a single day of protests as recently as June 19, 2014 (*Daily Amardesh* 2014). Civilians who kill protestors will go unpunished most of the cases when they are government supporters, whereas the victims' family members may face further harassment and death threats (Islam 2013). This government culture had begun since independence of Bangladesh in 1971 and is continuously deteriorating due to failure of democratic inclusion of the marginalized people in government system. Therefore, these forms of repression lead, in turn, to more extreme forms of frustration and resistance, including murder and political assassination. The government of India exploits security issues in Bangladesh to further its own interests, putting pressure on Bangladesh government to stop the "terrorists" (Hasib 2014).

Despite the repressive political climate in Bangladesh, there is also an entrenched culture of resistance that manifests itself in diverse ways. In the sections that follow I first discuss forms of resistance focused on resources and the environment. In the concluding section I focus on the broader movement in support of democratization and political inclusion so that the bottom-up development can protect livelihoods of the marginalized people. In both cases I attempt to document the ways that residents of Chapra in Kushtia District engage with these forms of resistance or, fail to engage, as part of a broader national phenomenon.

Struggles over resources and the environment

Chapra and other Ganges Basin communities in Bangladesh have deep-rooted, negative experiences of the Farakka diversion, the GK and GRRP. In this context, I first discuss the nature of the local and national protests that have been directed specifically against the Farakka Barrage and explain why the government of Bangladesh has not been able to find a remedy for this fundamental problem. I then describe the types of protests and actions that have arisen in response to more purely local events in Kushtia and nearby regions of Bangladesh.

Farakka protests

During one of our conversations, Tofajjal recalled his surprise at observing, as a fourth grade student in 1977, that the level of the Gorai River was at the lowest he had ever seen it and he asked his grandfather about this unusual condition. He was told that the Farakka Barrage was responsible for the water decline. Many people like Tofajjal, have similarly grown up understanding that their circumstances changed dramatically as a result of this single event. Joardar argued that the Farakka Barrage is their "death trap" and used the phrase *"vaate marbe; paanite marbe"* (they will kill us with rice crop damage and flooding) to express his feeling that the negative outcomes were deliberate, not incidental.

The pattern of resistance against the Farakka Barrage occurs within the historically developed relationship of India and Bangladesh, which has roots in previous periods of Bengali and Bangladeshi independence movements and the rupture of Bengal through the politicization of Hindu and Muslim religious identities. Keramot, Kamal and Ibrahim, focus group participants, analyzed the Farakka diversion in the context of Indian sub-continental politics and religion. They argued that the Hindu dominated India does not like Bangladesh. The British exploited India and Bangladesh under the Zamindari System they noted. Then Bangladesh was exploited by West Pakistan. Currently, as an independent country, Bangladesh is entrapped by India as a dominant, imperial power. They termed this domination as *dadagiree*, or elder brother's authoritative practices. Religion, they argued, performs a major role in the development of this political dynamics. The political party in India, Bharatiya Janata Party (BJP), has a Hindu nationalistic ambition and some political leaders of this party had negative attitudes toward Muslims. The BJP is in power from 2014 to current who was accused of killing for more than 1,100 Muslims in 2001 and is currently promoting new political dynamics. Moreover, influential members of the Indian

National Congress believe that the Hindu civilization was damaged by past Muslim Rulers and they need to revive Hindu culture through political mobilization (Daily Star 2013a). The Farakka Barrage was constructed under the Congress Government.

Currently, as an independent country, Bangladesh is surrounded by India on three sides and is dominated by it in a way that local people, males especially, refer to as dadagiree, a term that designates the way in which an elder brother traditionally exercises authority over younger brothers within the family. Dadagiree, as a feature of cultural identity, thus becomes part of what Sen (2008) calls the "political economy of power and inequality," noting also that "the coupling between cultural identity and poverty increases the significance of inequality and may contribute to violence" (Sen 2008:5). India plays the role of authoritative elder brother in this setting, not because it was first-born but because it is by far the more powerful country. The fact that India is predominantly Hindu while Bangladesh is predominantly Muslim reinforces India's sense of superiority.

The construction and operation of the Farakka Barrage is thus just one expression of how elder brother, India, behaves towards younger brother, Bangladesh. India's plans to build dams on other major rivers upstream from Bangladesh, repeat the pattern. India is building the Tipamukh Dam, for instance, on the Barak River in Manipur, an Indian state located on the eastern border of Bangladesh. India has also been implementing its National River Linking Project, which will move massive amounts of water from the Brahmaputra River, upstream from Bangladesh to the north, to desert prone provinces in India as far away as Rajasthan. These projects will cause ecological disasters for the rest of Bangladesh and have given rise to dharmaghats and demonstrations but the government of India dismisses these protests as acts of Muslim fundamentalists and takes various measures to stop the protestors.

India also acts as elder brother by marginalizing Bangladesh internationally and interfering in its internal affairs, including national elections. In 2013, for instance, then Indian External Affairs Minister publicly requested United States support for its position on pre-election political unrest in Bangladesh, arguing "that India's understanding of Bangladesh should be helpful to the U.S." (Dikshit 2013). This public statement followed others in which senior Indian government spokespeople termed the political protests, demonstrations and dharmaghats in Bangladesh as terrorist acts committed by Muslim fundamentalists (New Indian Express 2014) and suggested that, if the U.S. cooperated with India to stop the protestors, it would be an important contribution to South Asian anti-terrorist efforts. In this particular case, the US government did not support India's position, at least not publically, but India does have political leverage with the US and exercises it regularly in regard to Bangladesh. India openly supports pro-Indian movements in Bangladesh by every possible way. (Fox 2012; Times of India 2013). India is also increasing its military and para-military presence and extending its security fence along the Bangladesh border. The

Indian Border Security Force is reported to have killed 1,047 Bangladeshi people over the last 12 years (New Age 2014). In 2012, the security force stripped and tortured a Bangladeshi cattle businessman, Habibur Rahman, when he refused to pay bribes. The current government in Bangladesh rarely concerns itself with these forms of victimization. For example, when asked by the media about the Rahman incident, the Local Government for Rural Development Minister in Bangladesh stated publicly that, "the state is not too much concerned about it" (Daily Star 2012).

India has long supported the Bangladesh Awami League, one of two dominant political parties in the country and helped to ensure its long-term electoral success. As a result of this support, the Awami League (AL) is generally represented as "pro-Indian" (New Age 2013), whereas the leading opposition party, the Bangladesh Nationalist Party (BNP), is considered more likely to oppose Indian domination. Partly because of India's influence, the AL has ruled the country continuously since 2008 and they were also the party in power when the Ganges River Treaty was signed in 1996. As described in chapter two, the Ganges River Treaty, though usually referred to as a "water-sharing treaty," allows India to withdraw as much as 70 percent of Ganges River water during dry season. Opposition to the Farakka Barrage has nevertheless been led by leaders from both major parties. In 1976, when the dam began to operate, the Bangladesh Nationalist Party (BNP), led by President Ziaur Rahman, sought international assistance to resolve the Farakka dispute (Hossain 1981), an initiative that was vigorously opposed by the government of India. President Zia was assassinated in 1981, for reasons not directly connected to the dam dispute but no major political leader since then has taken as strong a position on this issue. The National Awami Party (NAP), then under the leadership of Abdul Hamid Khan Bhashani, also organized several protests and demonstrations against the Farakka Barrage in 1976. Many people from Chapra were actively involved in these movements. They attended the anti-Farakka long-march in May 1976, led by Bhashani, to demand the demolition of the dam. Bhashani died just a few months later, in November 1976.

Subsequently, in 1993, the International Farakka Committee (IFC), a major international organization, was organized to promote local voices against the effects of the Farakka diversion. The IFC has branches in several countries, including the United States of America, Canada and Sweden. The IFC organizes seminars, meetings, photo exhibitions, video and film shows, and demonstrations in Bangladesh and abroad to represent local voices at the national and international levels (International Farakka Committee 2011). The main objective of these movements is to put pressure on the governments in India and Bangladesh to develop proper water resource management for the Ganges Basin, to restore ecological systems and to protect the human rights of local people.

Little has changed, however, since 1976, and many local people are moving to more extreme forms of resistance because they are no longer optimistic about their futures. Many say that they do not care whether they live or die. Their own understanding about the Farakka domination, ecological vulnerabilities and livelihood failures are now based on media reports as well as traditional social networks and personal experiences. Many of them sit down at tea stalls to listen to the daily news on radios and TVs. Rickshaw pullers sit down on rickshaws and day laborers stand before the tea stalls. Based on this news and their own discussions, they develop critical thinking about the Bengali national politics and its domination over everyday livelihoods. As boatmen drive boats, they listen and participate in passengers' discussions. Farmers in their fields and women at cooking places share their pains and consider what they might do to bring improvement. Children are also active participants in these everyday forms of resistances. In some cases, local people generate conflicts between Muslims and Hindus as a backlash against Indian domination. For example, an idol of a Hindu Goddess and a Hindu temple, including two Hindu houses, were damaged in Kushtia on March 2, 2013 (Ghosh 2013). Political parties play blame-games about this destruction. The government of India and the Bangladesh Awami League blame the Jamaat-e-islami party and the BNP. None of these parties are willing to explore the root causes of these activities.

Local people encounter major frustration because of politicizing their water concerns. Ninety percent of the focus group respondents firmly believe that the government does not put enough effort into asserting their rights in regards to the Ganges flow concerns. But they do not want politicization of the Farakka issue to sway a national election. Every political party claims that they are committed to ensuring Ganges Basin water-sharing but they act differently when in power. Focus group respondents argued that they elect a new government every five years based on new commitments to resolve the Farakka dispute but their problems continue to increase. Keramot and Jalil want to see the emergence of new political leaders like Bhashani, who led the anti-Farakka long-march in 1976.

The current culture of historically developed power structures fails to recognize voices of local people like Jalil. The government follows India's bilateral preference: for example, the two treaties in 1977 and 1996 and two Memoranda of Understandings in 1983 and 1985 to share the Ganges flow. Because of this preference, the 1996 Ganges Treaty, the most recent treaty, fails to resolve water crises during the dry season (Rahman 2016). The government in Bangladesh fails to arrange a Joint River Commission meeting in 2010–16 because of India's disagreement (*Inquilab* 2016), although it was required to call emergency meetings according to the treaty. Despite this failure, the government in Bangladesh had blocked 18 rivers and canals in Bangladesh to facilitate transportation of the heavy loaded 140-wheel vehicles from India's mainland to northeastern states (New Age 2012a). Consequently, local people encounter major negative effects that are responsible for multiple reactions.

Kamal, Rahim, Fajal and Ibrahim argued that local movements take new turn against the domination of India inside Bangladesh. The governments in India are successful in exploiting the movements in Bangladesh in coordination with the global conceptualization of terrorism. They describe some movements of ecologically displaced people as the activities of Muslim fundamentalists and seek international supports against the global threats of terrorisms in Bangladesh. In 2013, for example, the Governments of India provided open patronization to the "pro-Indian movement" in Dhaka, the capital city (Times of India 2013). This political dynamics sometimes functions as one determining factor for the Ganges Basin decisions. To understand effects and frustration of India's management approach, it is important to focus here on local level movements.

Local protests

Rahim, Billal, Kamal and Jalil have been displaced repeatedly from their ancestral homes due to river bank erosion and embankment failures caused by development projects like the GK project and the GRRP. Many of them have been displaced permanently from local areas due to loss of croplands and houses. As a result of these types of hardships that have continued over two generations many local people are losing hope and becoming cruel and unsympathetic people who care little about their lives. More than one focus group respondent acknowledged that local actions are not limited to protests and demonstrations but extend to actions like the kidnapping of local elites or project staff and the stealing or destroying of project equipment.

The affected people boycott political leaders who snatch ballot box or support the snatchers. They want political leaders who have *shikor,* or root in community and in supports of the marginalized people. This pro-community approach can be understood with "moral economy of risk control" (Migdal 1996; Scott 1979). Based on this approach, they want political leaders who fight against criminalized governments and are available for them during the emergencies of flood or river bank erosion. As major supports for this objective, they raise consciousness by the different mechanisms like writing folk songs for their goal, meetings with fellow people, and demonstrations against the vote snatchers. Folk songs glorify honest leadership and describe negative effects of criminal activities. Local people stage roadside dramas with political content. Similar content finds expression during festivals and community feasts, as well as at meetings, demonstrations and rallies. Representatives of local people also raise consciousness against corrupt political leaders by national TV channels, for example, the drama series known as *420* ridicules and satirizes leaders. These discourses are carried to fields, kitchens and other work sites throughout the country and to all the places where people gather to converse and gossip. Although this discourse is often expressed within a hidden transcript, it serves as a form of "political mobilization" as well (Keefer and Khemani 2004) and informs both violent actions and more organized forms of resistance.

Local elites are becoming the subject of numerous strategies of revenge and vandalism, as well as political resistance. People spread rumors intended to bring competing elites into conflict with one another, usually on the basis of their political party affiliations. The goal is to weaken these elites or remove them from the local area. On occasion they destroy corrupted political leaders' local assets like fisheries projects and agricultural production materials like fertilizer. Many of them steal fish or poultry from the local elite to overcome starvation. Forty-two percent of survey respondents report taking water illegally from GK Project by breaking canals or building underground tunnels. They do this in such a way that nobody can trace them, performing these actions at night or during prayer times when most people will be attending a prayer house. Other marginalized people may know who is responsible but no one betrays the others since they all have similar frustrations.

In some cases, however, resistance to local elites occurs through acts of direct violence. On a few occasions, corrupt local political leaders have been killed. On November 13, 2010, for example, the ruling government's Member of Parliament (MP), Afazuddin, was attacked by a suicide bomber at his guest room in Kushtia (bdnews24 2010). In 2011, some unidentified people brutally murdered the municipality mayor and Awami League leader of Narsindi District (Daily Star 2011). Members of one extremist group, the Purba Bangla Communist Party, killed six people on April 24, 2003 to resist land grabbing and control over a fisheries project in Naogaon (Daily Ittefaq 2013). The fact that some of these violent actions are carried out by radical political organizations lends credibility to government claims that "terrorists" are responsible for all such acts of violence but the fieldwork evidence indicates that these organizations are responsible for only a small portion of the violence.

Resistance is by no means limited to these radical actions, however. The people most at risk of Gorai River bank erosion have submitted several memoranda to local governments requesting that they take steps against erosion and embankment failures. They follow a particular process to submit these memoranda. Community members discuss common issues like erosion during work in fields and cooking places and raise their concerns to community leaders. When the community leaders are available, they sit down together and discuss strategies to promote their concerns at the government level. They develop a one- or two-page draft of their demands and finalize it with participants' consent. Some people take responsibility to collect signatures and inform community members when they are going to submit this memorandum. A few community members will go together to submit the memorandum but all are concerned to avoid the local elites most closely associated with the government.

People affected by the GRRP have also submitted complaints to local governments against commercial usages of dredged sand when it interferes with their ability to use farmland or contributes to erosion and embankment failure. They have submitted memoranda to the Upazila Nirbahi Officer (UNO) at Kumarkhali Upazila and the Deputy Commission (DC) of Kushtia District. In addition, they have organized protests and demonstrations when local governments did not take necessary steps against illegal commercial activities. They organize and discuss these actions during gossip at tea stalls, while working in fields and during informal gatherings, as well as during more formally organized meetings.

They choose a protest site carefully, in order to attract as much government attention as possible. They gather, for instance, at the local district or sub-district government office. They have also cordoned off a dredged area to show government the evidence of destruction. Currently, the displaced people are desperate enough to initiate "do or die" actions and to sacrifice their lives for this cause because they have nothing more to lose. Fajal and Raju argued that these movements are their *baacha-morar lorai,* or survival-death struggles. If they are successful, they will get livelihood security; if they fail to get success, they will die because of starvation and malnutrition.

The pattern of resistance in Kushtia thus features everyday acts of resistance, peaceful petitions and demonstrations, and a growing number of violent attacks on elites staged by individuals and small groups of which only a small percentage belong to a known organization. No enduring local movements focused on resource issues have arisen in Kushtia, unlike in some other regions of the country. In Khulna District further south, local people raised their voices against the Khulna Coastal Embankment Rehabilitation Project (KCERP) and developed local movements called Beel Dakatia movements (Dakatia Wetland movements). In 1990, they broke four KCERP embankments to allow standing water to drain away from fields, thus restoring agricultural production and employment and protecting themselves from the health hazards associated with stagnant water (Unnayan Onneshan 2006:5). Their main complaints are that the government is ignoring their voices about local ecosystems in the southeast regions. Because of the Beel Dakatia Movements, the Asian Development Bank was forced to cancel the KCERP project and develop a new project, the Khulna-Jessore Drainage Rehabilitation Project (KJDRP); however, local people did not accept the KJDRP either because of their concern over losing the Hamkura River (Islam and Kibria 2006:17). Local people formed the Hamkura River Bachao Andolon (Hamkura River Protection Movement) and forced the authorities to cancel the embankment project (Islam and Kibria 2006:18). Victims of KCERP and KIDRP projects sued the government for compensation with the assistance of the Bangladesh Legal Aid and Services Trust and the Bangladesh Environmental Lawyers' Association (Unnayan Onneshan 2006:8).

A local movement also emerged in the southwest region of the country against the Bhabadha Sluice Gate and Kashimpur Embankment Project. In this case, local law enforcement agencies launched legal actions against the protestors (Islam and Kibria 2006:20). More recently, local movements, known as Beel Kapalia (Kapalia Wetlands) movements, also emerged in 2012 in opposition to the Tidal River Management (TRM) project in Jessore District, to the south of Kushtia. The Member of Parliament (MP) of Jessore-6 parliamentary constituency, along with his associates in local governments, encountered extreme resistance when they went to inaugurate the TRM in 2012. The project goal was to raise the wetland soil level to overcome water stagnation and improve agricultural production. However, local people of Manirampur in Jessore held different opinions about TRM. According to their spokespeople, this project will decrease the existing cropping area in the wetlands. The MP was unable to inaugurate the project and left the area after he and local upazila chairman were wounded and transferred to Dhaka for medical treatment. The agitated demonstrators also burned the vehicles of the MP and his associates. They continued sit-in demonstrations and threatened to die by poison until the government accepted their demands (New Age 2012b). In 2016, more recently, for example, local people were raising their concerns over the coal-fired power plant at Gondomara Union, Banskhali Upazila in Chittagong District and engaged in conflict with law enforcing agencies which were killed 4 villagers, wounded 30 including 11 police (Hussain 2016).

The major reasons behind these movements are to reduce their concerns over livelihood failures including forced migration and displacements. Between 2001 and 2011, disasters caused approximately two million people to migrate within Bangladesh (T.J. 2013) and many others have left the country (Swain 1996). Many of those displaced within the country end up in the slums of Dhaka, living under conditions no better or worse than those they left.

Consequently, many of the local movements are sometimes directed towards abolishing large-scale water infrastructures and restoring the ecosystems on which there were formerly able to depend on their livelihoods. They also want to promote ecological health and siltation for cropland fertility based on the Gorai flow. Although no enduring social movements of this type have emerged in Kushtia, local people in Chapra itself attempted to destroy an embankment on the Gorai River in 2011. Forty-six percent of survey respondents report wanting the embankment removed. Their objective is to retain the wild river fisheries that need the regular Gorai flow to continue and, currently, the embankment interferes with this flow. An elite family and their allies oppose removal of the embankment because they have created a fish pond behind it, using the flood waters that overflow the embankment during the monsoon season. They can also pump water from the pond to their croplands during certain times of the year. The embankment and fish pond both interfere with the hydrology of the area and the pond becomes a vector for illness during times when there is little rainfall and the water becomes stagnant. The families that created the pond have placed guards on the embankment and used their connections to local government in an attempt to resolve the conflict but the movement against the embankment continues.

Localized movements organized by national organizations are also occurring frequently throughout the country. The Policy Research for Development Alternatives is helping to organize local opposition to HYV crops like Nerica and Bt Brinjal, because of the threat they pose to local biodiversity and food self-sufficiency. This organization promotes new farmers' movements by organizing local farmers and raising consciousness based on their research findings. In 2010, for example, more than 500 farmers from 19 districts in Bangladesh gathered in a meeting in Tangail to raise their voices against new agricultural technologies. According to them, their agro-ecological experiences, local knowledge, local seeds and natural fertilizers can ensure food sovereignty but the new technologies cannot. They further argued that hunger and climate change challenges in Bangladesh can best be addressed by continuing to use the 15,000 local paddy varieties (UBINIG 2011). The Policy Research for Development Alternatives also promotes dialogue among grassroots farmers and ministers, MPs, scientists and government officials to reduce the gap between local and scientific knowledge. This dialogue can reduce the gap when democratic system in Bangladesh is fully functional meaningfully.

Struggles for democratization

Ninety percent of the total population of Bangladesh can be categorized as marginalized in terms of their economic status and level of exclusion from political institutions. According to Nuruzzaman (2007), 45 influential families control the national economy of Bangladesh. The Prime Minister (PM), Sheikh Hasina, belongs to one of these elite families, along with other senior members of the governing Awami League, the party she leads. Former Prime Minister, Khaleda Zia, leader of another dynastic party in Bangladesh, Bangladesh Nationalist Party (BNP), also belongs to one of Bangladesh's elite families. Government decisions, no matter what party is in power, are often connected to business interests and projects, such as the GRRP, provide enormous opportunities for elite families and their allies to make money legally and illegally. This pattern of elite domination and corruption appears to have been expanding over recent years. For example, the average income from all sources of key figures with the Bangladesh Parliament (the Deputy Leader, Chief Whip and other Whips) was reported to have increased 15,281 percent from 2009 to 2013 (UNBconnect 2013). Government process is thus characterized by a lack of transparency, by corruption and patronage and an extraordinary degree of hierarchical control at the top end of the system. Consequently, according to one of my focus group respondents, Kusum, "*rajnoitik-der proverb-er karoena amora shubidabonchito manusgulo aaj ardomrita*" (we, the deprived people, are almost death because of this political domination.)

It is not surprising then that a large number of organizations have sprung into action with the objective of abolishing this hierarchical chain of command and promoting "true democracy" (Blair 1996; Petreas and Velmeyer 2012). As noted previously, the government responds to these organizations by criminalizing approach using law enforcement agencies to repress them and by developing its own cadre of political party supporters to intimidate, harass and even assassinate movement leaders (Pilger 2013). Party supporters who commit criminal activities are sometimes charged but will be protected by the government itself. For example, the Awami League government in 2009–2013 withdrew more than 7,131 criminal cases against their supporters (*Financial Express* 2012a). Political leaders deliberately and publicly encourage violent actions by their supporters with political statements such as: "*ekter bodole doshta lush chai*" (avenge one death among our supporters with ten deaths among our enemies) (Rehman 2013).

One comment from an ordinary citizen, eloquently expresses the frustration of the disenfranchised 90 percent. Geeta Sarkar, an ordinary citizen, was riding a bus on December 1, 2013, to a scheduled appointment during the time of a demonstration organized by an opposition party, BNP. During their efforts to block all movement of traffic, some unidentified "goons," hurled firebombs into the bus that burned 19 people. Two people died on the spot and 14 others suffered permanent disabilities, such as hand or leg amputations. The seriously injured people, including Geeta Sarker, were admitted to a hospital where the prime minister later visited to express sympathy. Rather than show appreciation to the PM for her concern, and thus be captured in the propaganda war between the government and the opposition parties, Geeta Sarkar expressed her frustrations against the existing political culture as follows: "*apnara amader niye chinimin khelchen. Amora valo sorkar chai, ashustha sorkar chaina*" (You, the Prime Minister and the opposition leaders, are exploiting us. We want a democratic government; we do not want criminalized governments (*Prothom Alo* 2013)). Geeta Sarkar thus identified a core issue in a way that appears to have captured the mood and frustrations of the country. Kusum, my focus group respondent, signaled new steps to reduce these frustrations: "*sadharan manuske boka bananor sportha bodhoy rajnitibid-der thaka thik noy*" (political leaders should not be dare to make us, general people, fools).

There are many political parties in Bangladesh other than the two major political parties that are locked in a dynastic political struggle and, some of these parties, if elected, might well govern in a more genuinely democratic way. Elections in Bangladesh are won, however, through vote rigging, buying votes, intimidation and assassination and many people therefore are frightened away from practicing their voting rights (*New Age* 2011). Local people sometimes boycott political leaders who practice these activities. They want political leaders who have *shikor*, a 'root' in the local community, and who are genuinely interested in helping people affected by ecological vulnerabilities. They want political leaders who fight against the criminalization of government. But the main expression of these aspirations does not appear to be directed toward the creation of alternative political parties so much as towards the articulation of a "hidden transcript" as Scott has described it. These discourses are carried to fields, kitchens and other work sites throughout the country and to all the places where people gather to converse and gossip. Although this discourse is often expressed within a hidden transcript, it serves as a form of "political mobilization" as well (Keefer and Khemani 2004) and informs both violent actions and more organized forms of resistance.

The number of human rights organizations in Bangladesh is expanding as the level of conflict escalates and peaceful remedies are rejected. Most local movements seeking social justice and government transparency and accountability begin with actions like demonstrations and the submission of complaints and petitions to local authorities. When these complaints are ignored or when the government responds in a violent way, levels of violence escalate on both sides. In 2013, more than 507 people were reported killed and 22,407 people injured in Bangladesh because of conflicts between law enforcement agencies and local people in Bangladesh (*Daily Star 2014*). Some of these deaths occurred in Kushtia where, for example, police killed a regional commander of an outlawed political party, the Gono Mukti Fouz (People's Freedom Force), on September 23 (Daily Star 2013c). Some masked men attacked the Upazila Nirbahi Officer with bombs at Bheramara in Kushtia (Daily Star 2013b). On August 28, 2012, some unidentified people killed a Union Council chairman and his two associates with firearms at Kumarkhali in Kushtia (*Financial Express* 2012b). The Rapid Action Battalion, an elite anti-crime and anti-terrorism police unit, killed one youth in Kushtia on April 5, 2013 (*Daily Amardesh* 2013). Violence reaches much higher levels in other regions of the country, however. Between 1972 and 2013, 75 UP Chairmen were killed because of political conflict in Khulan Division alone (Hossen 2014). State-sponsored abduction is reported to be another major mechanism to stop the protestors (Bergman 2013).

The recent proliferation of non-violent human rights and civil society organizations offers at least a glimmer of hope within the current climate of fear and hostility. Organizations like Citizen Rights Forum oppose both criminalized government systems and violent actions by protestors. The Bangladesh Society for the Enforcement of Human Rights (BSEHR) acts as advocate for oppressed people in a number of ways, through investigation, fact-finding and legal aid support, arbitration processes, educational programs, monitoring and policy development. According to their website information, the target groups of the BSHER are "the poor and marginalized helpless women, children, juveniles and male victims of violations of human rights" (Bangladesh Society for the Enforcement of Human Rights 2014). Another organization, the Legal Aid & Human Rights Organization (ASK) provides legal support to the victims of Rapid Action Battalion (RAB) and police harassment, torture, crossfire injuries, extrajudicial killings and abductions (Ain O Salish Kendra 2008). Bangladesh Legal Aid and Services Trust (BLAST) works in 19 districts, including Kushtia and Dhaka, where it is headquartered. A panel of lawyers provides legal support to marginalized women and children who are the victims of powerful people in the country. The major goal is to establish social justice, human rights, liberty, freedom of expression and equality (Bangladesh Legal Aid and Services Trust 2010).

Inclusion in government systems

The focus group respondents Rahim and Kafil want to be included in political leadership. Currently, they do not have scopes for raising their voices in political party organization as leaders practice *jee hujur,* or "yes sir," like an army chain of command. The failures to do so cause punishments like demotion or expulsion. The major cause behind this practice is that no political party practices proper democratic systems in selecting party officials. The major political party is operated with a single person, the party chief, who enjoys supreme authority from the central to local level. Sometimes, the party chief appoints local committee chairmen or members based on the criteria, for example, muscle power to fight with opposition groups. Based on this patronization, the appointed officials are busier to satisfy the party chief and personal interests and detached from communities. They want to gain more from government benefits like subsidized agricultural materials or water development projects like the GRRP. The focus group respondents do not want this criminalized party organization. Kamal, Raju and Kofil want to get democratic practices inside the party organizations. They want to get recognition of leadership criteria like honesty, education, popularity and accountability. One major organization, *SHUJAN* or Citizens for Good Governances, represents local voices for this purpose.

Local movements are to promote democratic systems by reducing malpractices of political parties and government systems. Currently, many leaders got involved in vote rigging, intimidation, killing, ballot-box and paper snatching to win in an election. Consequently, local voters fail to practice their voting rights (New Age 2011). Based on this election engineering, the FGD respondents Jalil and Keramot informed me that criminals get *ijara,* or formal control over local government institutions like Upazila Parishad (UP) or Union Council (UC) with a position of chairman or member and fulfill personal interests. These elected officials at local government justified their malpractices in the name of democratically elected representatives. They are busier to gain from ecological vulnerabilities like river bank erosion, embankment failure, flood, rainy season failure, water stagnation and drought. They sell subsidized tubewells, chemical fertilizers and HYV seeds for monetary gains. Consequently, the marginalized people encounter exclusion from these government services. Local movements seek to promote democratic practices inside political party, to fight against this exclusion, and to promote democratic values, inclusion and social justice.

Local people seek to incorporate their voices in local governments based on their representation. The focus group respondents Rahim, Kamal and Billal argued that local government staffs do not treat them with respect as they have least representation; therefore, they exclude from social service program benefits like food, housing and health care. To overcome this exclusion, they seek to use voting rights as a major mechanism for democratic election and representation. According to Billal and Joardar, political leaders need to count their voices for winning election because they are the majority of people at Chapra. They also argued that many of their movements are producing positive outcomes; some of their community leaders are successful in wining local government election as member or chairman of UC or UP. Their movements seek to get greater successes because they are majority of local people.

Because of their negative experiences, most of the focus group respondents do not supports any political party. They use voting rights based on the criteria of *monder valo* (relative good) and always put pressure to promote inclusion and democratic practices. However, they are facing major challenges from the criminalized government systems. The protestors against the corrupted practices of government describe as criminals; the government executes extrajudicial killings by law enforcing agencies like Rapid Action Battalion (RAB) and police (Amnesty International 2012). The elite force, RAB, received training from the United Kingdom (Deccan Herald 2011). Again, state-sponsored abduction is another major mechanism to stop the protestors (Bergman 2013). The affected people are failing to keep tolerance on these criminalized government systems and executing counter actions that are not suitable for civilized systems. The inclusion in government system emphasizes on reducing control of local elites over local development projects like the GK. Currently, local elite farmers control the project water for their own crop production, and therefore, the marginalized farmers are excluded from the project water supply which creates major reason for initiating revengeful actions against the elite farmers. They initiate the different types of illegal actions like aisle cut off to get water from the elite farmers' croplands. They create further harms for the rich farmers by transferring fertilized water. These activities can be termed as "weapons of the weak" to make their voices heard in the government system so that they can be recognized in the development projects. This realization of the success goes beyond Scott's focus as he never says the "weak" ever win – he just documented their resistance.

Movements for bottom-up development approach

Most of my focus group respondents would like to include their voices based on the bottom-up water and agricultural management approach. Currently, the head of government in Bangladesh, Prime Minister (PM), is a major decision maker for a major development project like the GRRP and other political leaders and agencies follow this decision without raising their concerns. This decision is well connected with the interests of businessmen that can be described as ruling power structure in Bangladesh. This ruling structure sometimes does not care about accountability; many of them are busier with personal interests. For example, total income of the immediate past mayors of Rajshahi, Khulna, Barisal and Sylhet city corporation has increased 6,747 percent during the past 5 years from 2009 to 2013 (Daily Star 2013a). Again, the government in Bangladesh spent more than US$300 million to restore the Gorai River although the river is almost disappeared. The focus group respondents at Chapra are raising their voices against this development domination and their displacements spread local movements at regional, national and international level. To overcome these negative effects, some organizations are raising their voices so that local people can protect their rights and responsibilities. For example, the Transparency International Bangladesh (TIB) is promoting local voices against corruption and malpractices of the government. The UBINIG, Policy Research for Development Alternatives, with *Noya Krishak Andolon* or new farmers' movements are also representing local voices to promote the bottom-up approach (Mazhar 1997). Based on the bottom-up development approach, local people seek to raise their voices over ecological vulnerabilities that originate from water and agricultural development failures. Their frustration against the GRRP is increasing due to increasing livelihood vulnerabilities like damage or destruction of household infrastructures, assets, croplands and field crops. Some major areas in Chapra, Mohanagar, Sukhdippur and Gopinathpur are currently encountering erosion due to negative effects of the GRRP dredging and other development activities. When local people express their concerns, the government treated them as troublemakers and took legal actions. Consequently the focus group respondents Joardar and Jalil do not have a specific shelter place to ensure their housing that adds new causes of local movements. The failures of livelihood components like housing make them desperate people to raise their voices to the government level.

Local people seek to promote their voices over water commodification and eco-agriculture challenges. The focus group respondents Kamal, Jalil, Rahim and Raju are frustrated because the government did not incorporate their voices about ecosystems to execute the GK project that decreases local self-sufficiency and increases agricultural commodification. The majority of the focus group respondents argued that they cannot cope with the commodification of the GK and GRRP projects. They are turning into consumers of rich farmers for agricultural commodity like fish, crop and irrigation water due to failures of common ecological resources. Rahim and Kamal termed the GK project, deep tube-well, power tiller and paddy husking machine as their *shotru*, or "enemies," because of reducing their employment opportunities and commodification. Their frustration extends further with the project mismanagement and seasonal irregularities. One in-depth case respondent Joardar argued that the GK project provides water earlier or later than the cropping schedule. Furthermore, Rahim and Fajal, like most of other focus group respondents, failed to get proper water supply in the higher and lower croplands because the project fails to recognize agro-ecological systems during the construction period. Therefore, 92 percent of people in Kushtia who are marginalized farmers encounter survival challenges. Among them, according to the household survey data, 42 percent of farmers initiate the different types of actions to get water supply from the project. They want to be included in the project's decision making and implementation process to raise their voices over the commodification and agro-ecological systems.

In legal term, locally established Water Management Organizations (WMOs) should perform this function but none is existed for this role in major water infrastructure projects. Part of this disconnect is the split in responsibilities between the financiers of these projects and the government agencies. Many of the staff in Bangladesh Water Development Board (BWDB) is strong supporters of participatory management based on their local water management experience which needs political government's supports. I think the real profound disconnect is the power of the dynastic parties and politicians who run as much of the system as they can for their own interests. Then there are good professional bureaucrats who face interventions, legislation, orders, etc., that continually defeat them. Therefore these personal interests create major reasons for local people's sufferings and movements for participation and representation.

Based on this bottom-up approach, local people would like to overcome environmental degradation that originates from effects of water and agriculture development projects. Kamal, Kusum, Keramot, Kafil and Jalil argued that the GK project infrastructures are responsible for water stagnation and agro eco-system failures. They seek to promote their understanding about local fisheries and causes of their extinction. They want to raise their voices about fertilizer and pesticide contamination. Domestic animals encounter diseases and death from chemical fertilizers and pest contamination in grazing lands. They are not responsible for these socioecological problems although they face major negative effects. They are very frustrated with these problems and want to get back ecosystems to protect livelihoods.

Only the bottom-up water and agricultural development approach can protect local ecological knowledge and can preserve their past agricultural skills and practices. The focus group respondents express frustration because of continuous ecological knowledge reduction for agricultural production due to the Ganges flow reduction and technological domination. Their children fail to learn this knowledge from older generations that generate major concerns of losing *shikor,* or knowledge root. The marginalized groups like women lost traditional income opportunities from fishing, paddy husking, water collection, traditional irriga-tion, fodder collection and domestic animal that deteriorate household economic conditions. Fishermen and boatmen encounter unemployment due to the basin flow irregularities and commercial fish production. Their earnings from boat driving are replaced with mechanized boat and many of them encounter unem-ployment and exploitation because of this commercial sector. They fail to perform responsibilities as fathers, mothers, sons, daughters or relatives that create frus-tration and grievance against the factors responsible for these sufferings. Billal experiences his children's starvation and his wife looks at him with full of frustration and helplessness. They want to protect linkages between agroeco-systems, indigenous seed production and traditional fertilizer practices based on local knowledge and traditional employment opportunities. For this purpose, they are raising their voices so that they can be represented in local and central governments.

The goal of local movements pursues to overcome the displacements of marginalized people. Rahim, Billal, Kamal and Jalil, encountered displacements for more than one time due to execution of development projects like the GK and GRRP. Many of them displaced from their ancestral homes permanently due to lose of croplands and houses. They encounter these sufferings over generations, and currently, many of them have nothing more to maintain decent livelihoods. Rahim and Kamal argued that they do not have much difference between life and death because of human rights failures of food, health care and housing. These failures make them desperate people against the responsible people, development projects and infrastructures which cause multiple forms of local movements.

Many of them carry firearms to achieve their objectives. For example, law enforcement agencies are reported to have arrested a person for possession of illegal firearms in Kushtia on January 21, 2013 (Mehtab 2013). Some of these illegal actors are truly criminal in nature, rather than political, as in the case of the thieves who killed a woman with a bomb while stealing at Daulatpur in Kushtia on December 17, 2012 (Saky 2012). Bombs as well as firearms are routinely used in smuggling operations as well and, on June 6, 2014, the police were reported to have arrested a man along with a fire arm and three bombs at Kumarkhali in Kushtia (Rosul 2013). Those who resort to these actions themselves experience arrest, torture, kidnapping and extrajudicial killings by law enforcement agencies and ruling political leaders (UNBconnect 2014). Despite the visible presence of legal aspects of these social problems, it is important to emphasize more on socioecological aspects as the root causes.

Local movements pursue freedom from elite domination over social services: such as, food, employment, education, health care and housing. Some of these projects are Food For Work (FFW), Employment Generation for the Ultra Poor (EGPP), School Feeding Program, Revitalization of Community Health Care Initiative and Maternity Allowance Program and *Gucchagram,* or slum, that describe social service programs of food, employment, education, health care and housing, respectively. They face major frustration when the deserving people are excluded and the non-deserving people are included for these benefits. Because of this exclusion, their children face illiteracy although their due benefits are exploited for personal gains. Some of the major effects of these failures are illiteracy, malnutrition, maternal death, housing failure and displacement; therefore, they are front side activists of local movements. The affected people like Kafil lose patience when the starved people or beggars visit them for food and beg because of exclusion. The displaced people live in informal places and develop their own forms of movements against elites. They want to raise their voices over local social service management committees based on their representation. They want inclusion on the different committees like irrigation management, seed and fertilizer management, food security, employment, education and housing. For this purpose, they want to elect community friendly MP and UP and UC chairmen and members.

Moreover, local people seek the types of social services fit with local socioecological conditions. Currently, the central government makes decision about types and amount of social services for the marginalized groups of people. Billal and Rahim do not want the school feeding program that is not helpful for their socioeconomic conditions as they cannot send their children to school. Fajal and Keramot do not want the slum program that creates isolation and stigmatization from neighbors. They are getting extremely frustrated with health care programs' commodification. Therefore, some of them destroy educational infrastructures based on their painful memories of dropping out from school due to education fund mismanagements. They destroy health infrastructures based on the grievance of their health service failures and experiences of abusive language and scolding from health service providers. Transportation like private cars or bus is also major targets of vandalism and burning because, according to Abonti, Fazlu, Suman and Sabbir, transports represent symbols of corruption. The affected people resist corrupted people from their neighborhoods based on their past memories of suffering, harassment and exploitation. These natures of local movements can be described with Hobsbawm's (1959) explanation of culture and organization in the context of social movements in Europe.

Conclusion

Local movements seek to restore ecosystems, overcome ecocracies and socio-ecological vulnerabilities. In this context, the affected people at Chapra have deep rooted grievances against the Farakka Barrage, GRRP and GK Project. They encounter irregular basin flow during the rainy and summer seasons that create major survival challenges for many of the marginalized people at Chapra. Their challenges turn in extreme level when they lose everything like field and stored crops, production materials, croplands and household infrastructures because of these irregular seasonal flows and river bank erosion or flood. The government in Bangladesh fails to protect the basin water rights due to India's ignorance of ecosystem. The level of their concerns makes them desperate people because their livelihood sufferings do not have difference between life and death. This frustration expresses in local culture and daily forms of movements. They do not want to tolerate political leaders who control their social service benefits. The failures to get these benefits from education, health care and housing programs turn them into extremely vulnerable position. They are desperate to resist political leaders who make business based on their unemployment, malnutrition, sickness and illiteracy. Their movements are described as acts of opposition political party to destabilize governments; this description provides scopes for legal action of law enforcing agencies. Sometimes, extrajudicial killings stop protestors that invoke new level of movements. The affected people lose temperaments and get involved in the different violent activities like vandalism, burning and killing of government infrastructures and some of them do not hesitate to kill the corrupted people.

The governments in India and Bangladesh need to recognize the root causes of this level of local movements before it is getting too late. This realization is not only important for the basin communities, but also for protecting the government systems. In this context, the governments of India need to recognize importance of the basin-wide water governance for ecological integrity and community livelihoods. The government of Bangladesh needs to recognize importance of bottom-up water and agriculture policies and implementation. Local movements seek to promote community voices over water and agricultural policy making and implementing process. Based on this bottom-up approach, they seek to resist discrimination and social inequality and to promote equality and social justice. They further want to change criminalized political systems that describe their movements as criminal activities and increase repression and harassment. In this context, they want democratic practices in political party leadership so that they can be included in government systems. This inclusion can uphold their voices of livelihood practices and human rights with proper water policy and governance.

Some organizations are working for representing local voices over policy making and implementation processes. The activities of these organizations are very limited by comparison to the scale of the problem, which is concentrated in the capital, Dhaka. However, they provide evidence of the possibility of a broader and more all-encompassing movement away from political violence and they are helping to lay the groundwork for the changes I recommend in my next and concluding chapter. The human rights, on which I focus in my study of Chapra, to water, food, employment, education, health and housing, cannot be restored on the basis of purely regional reforms to the water governance system. Reform will need to occur on a national level as well. It would be naive to think that such changes are imminent or that the government will move towards wholesale reforms voluntarily. However, it is increasingly clear that Bangladesh is reaching a crisis point and behaving more and more like a failed state. It is at least possible that senior government leaders will realize, before it is too late, that some measure of reform is necessary, if only to protect their own lives and self-interest.

References

Ain O Salish Kendra (ASK) 2008. Keeping Promises, Defending Beliefs, Protesting Injustice & Redressing Inequality. http://www.askbd.org/web/?page_id=672, accessed January 1, 2013.

Amnesty International 2012. Annual Report 2012; The State of the World's Human Rights-Bangladesh. http://www.amnesty.org/en/region/bangladesh/report-2012#section-12–3, accessed May 14, 2012.

Bangladesh Legal Aid and Services Trust (BLAST) 2010. Who We Are. http://www.blast.org.bd/who, accessed July 25, 2013.

Bangladesh Society for the Enforcement of Human Rights (BSEHR) 2014. Bangladesh Manabadhikar Bastayan Sangstha. http://www.rtiforum.org.bd/index.php?option=com_comprofiler&task=userprofile&user=132&Itemid=13, accessed June 21, 2013.

Bdnews34 2010. I Held for Kushtia Suicide Blast. http://bdnews24.com/bangladesh/2010/11/20/1-held-for-kushtia-suicide-blast, accessed November 21, 2012.

Bergman, D. 2013. War Criminal Trial: Witness Alleges State Abduction. http://www.newagebd.com/detail.php?date=2013–05–16&nid=49319#.U7Be7u_QfIU, accessed May 17, 2013.

Blair, H. W. 1996. Democracy, Equity and Common Property Resource Management in the Indian Subcontinent. *Development and Change* 27(3): 475–499.

Daily Amardesh 2013. 2013. RAB Killed One Person at Kushtia, trans. http://www.amardeshonline.com/pages/details/2013/05/04/198658#.U7Bdw-_QfIU, accessed May 5, 2013.

——— 2014. 15 Murders a Day in the Country, trans. http://www.amardeshonline.com/pages/details/2014/06/20/247224#.U7BdZu_QfIU, accessed June 21, 2014.

Daily Inquilab 2016. JRC Meeting is in Limbo for Six Years Due to India's Disagreement: No Water in River. January 21. accessed April 19, 2016.

Daily Ittefaq 2013. 46 People Gets Life-long Sentence for Killing Six People. http://www.ittefaq.com.bd/index.php?ref=MjBfMDVfMTZfMTNfMV8xXzFfNDEw MTg=, accessed May 25, 2013.

Daily Star 2011. Narshindi Mayor Shot Dead. http://archive.thedailystar.net/new Design/news-details.php?nid=208955, accessed June 21, 2012.

——— 2012. Pranab, Ashraf Play It Down. http://archive.thedailystar.net/new Design/news-details.php?nid=219359, accessed May 12, 2013.

——— 2013a. Wealth, Income Soar in 4 Years. http://archive.thedailystar.net/beta2/news/wealth-income-soar-in-4-years/, accessed June 15, 2013.

——— 2013b. Bomb Attack on UNO in Kushtia. http://archive.thedailystar.net/newDesign/cache/cached-news-details-270569.html, accessed February 23, 2014.

——— 2013c. Outlaw Killed in 'Gunfight' with Kushtia Police. http://archive.thedailystar.net/beta2/news/outlaw-killed-in-gunfight-with-kushtia-police/, accessed January 3, 2014.

——— 2014. Rights Situation Was Alarming in 2013, Says Rights Body ASK's Report. http://www.thedailystar.net/rights-situation-was-alarming-in-2013-4883, accessed January 5.

Deccan Herald, 2011. UK Defends Training for Bangladesh's Elite Force RAB. http://www.deccanherald.com/content/135955/uk-defends-training-bangladeshs-elite.html, accessed May 21, 2013.

Dikshit, S. 2013. India's Understanding of Bangladesh Will Help U.S. http://www.thehindu.com/news/national/indias-understanding-of-bangladesh-will-help-us/article5516435.ece, January 5, 2015.

Escobar, A. 1992. Reflection on 'Development:' Grassroots Approaches and Alternative Politics in the Third World. *Futures* 0016–3287/05411–16.

Financial Express 2012a. BNP to Revive All Withdrawn Cases. Dhaka. http://www.thefinancialexpress-bd.com/old/more.php?date=2012–02–02&news_id=96765, accessed February 5, 2012.

Fox, J. 2012. The Bengal Tigers in the R&AW Cage: R&AW Trained Crusader 100 in Action in Bangladesh. *Sri Lankan Guardian*. April 23.

——— 2012b. UP Chairman Among 3 Gunned Down in Kushtia. http://www.thefinancialexpress-bd.com/old/more.php?news_id=141493&date=2012–08–29, accessed June 2, 2012.

Gadgil, M. and R. Guha 1992. *This Fissured Land: An Ecological History of India*. Berkeley: University of California Press.

Ghosh, R. 2013. Monthly Report-March 2013. http://www.google.ca/url?sa=t&rct=j&q=&esrc=s&source=web&cd=1&ved=0CB4QFjAA&url=http%3A%2F%2Fhrcbmdfw.org%2Ffiles%2F862%2Fdownload.aspx&ei=-6KvU5LgPNXqoASehIKIBg&usg=AFQjCNHY_BSL50D6WIyiPMxlrJJ_vgAdNw&bvm=bv.69837884,d.cGU, accessed February 28, 2014.

Hasib, N. I. 2014. India Influences Bangladesh Policies: US Report. http://bdnews24.com/bangladesh/2014/05/01/india-influences-bangladesh-policies-us-report, accessed May 5, 2014.

Hobsbawm, J. 1959. *Primitive Rebels: Studies in Archaic Forms of Social Movements in the 19th and 20th Centuries.* Oxford: The University Press.

Hossain, I. 1981. Bangladesh-India Relations: Issues and Problems. *Asian Survey* 21(11): 1115–1128.

Hossen, M. H. 2014. 1971 Persons Killed in Khulna Division in the Last Five Years, trans. http://www.jugantor.com/news/2014/01/05/56098, accessed January 7, 2014.

Hussain, A. 2016. Banskhali Locals Vow to Protect Land. http://www.dhakatribune.com/bangladesh/2016/apr/06/banshkhali-locals-vow-protect-land, accessed April 14, 2016.

International Farakka Committee (IFC) 2011. The Activities of International Farakka Committee. http://www.farakkacommittee.com/IFCactivities.php, accessed May 13, 2013.

Islam, M. S. ed. 2013. *Human Rights and Governance Bangladesh.* Hong Kong: Asian Legal Resource Centre.

Islam, S. and Z. Kibria 2006. *Unraveling KJDRP: ADB Financed Project of Mass Destruction in Southwest Coastal Region of Bangladesh.* Dhaka: Uttaran.

Kapoor, D. 2011. Adult Learning in Political (un-civil) Society: Anti-Colonial Subaltern Social Movement (SSM) Pedagogies of Place. *Studies in the Education of Adults* 43(2): 128–146.

Keefer, P. and S. Khemani 2004. Why Do the Poor Receive Poor Services? *Economic and Political Weekly* 39(9): 935–943.

Khan, T. R. 1996. Managing and Sharing of the Ganges. Natural Resources Journal 36: 456–479.

Mazhar, F. 1997. NayakrishiAndalon: An Initiative of the Bangladesh Peasants for Better Living. http://www.idrc.ca/cp/ev-85301–201–1-DO_TOPIC.html, accessed May 29, 2013.

Mehtab, T. B. 2013. *UNO Injured in Kushtia Bomb Attack.* Dhaka: Bdnews24.

Migdal, Joel S. 1996. *Peasants, Politics, and Revolution: Pressures Towards Political and Social Change in the Third World.* Princeton, NJ: Princeton University Press.

Mithu, A. I. 2011. So Many Hartals? http://archive.today/cnRrG, accessed August 26, 2012.

New Age 2011. One Killed, Ballot Boxes, and Paper Snatched. http://newagebd.com/newspaper1/archive_details.php?date=2011–06–04&nid=21290, accessed July 1, 2012.

―――― 2012a. 2 Students of Kushtia Medical College Drown. http://newagebd.com/detail.php?date=2012–07–28&nid=18706#.U7CMlO_QfIU, accessed September 25, 2012.

―――― 2012b. Stop Harassment of Farmers, Bhabadah Movement Committee. http://newagebd.com/detail.php?date=2012–06–05&nid=12733#.U6-PdO_QfIU, August 11, 2012.

―――― 2013. Ruling Party's India Game and Bangladesh's Shame. http://newagebd.com/detail.php?date=2013–12–06&nid=75624#.U6-PL-_QfIV, accessed April 21, 2013.

―――― 2014. Lives Lost in Vain. http://newagebd.net/1759/lives-lost-in-vain/#sthash.Yziw27Mm.dpbs, accessed May 1, 2014.

New Indian Express 2014. Support Democracy and Secularism in Bangladesh. http://www.newindianexpress.com/editorials/Support-Democracy-and-Secularism-in-Bangladesh/2014/01/07/article1986288.ece, January 27, 2014.

Nuruzzaman, M. 2007. Neoliberal Economic Policies, the Rich and the Poor in Bangladesh. *Journal of Contemporary Asia* 34(1): 33–54.

Petreas, J. and H. Velmeyer 2012. Imperialism and Democracy: Convergence and Divergence. *Journal of Contemporary Asia* 42(2): 298–307.

Pilger, J. 2013. The Prison That Is Bangladesh. http://www.theguardian.com/commentisfree/2013/dec/15/prison-bangladesh-moudud-ahmed, accessed March 13, 2014.

Prothom Alo 2013. We Do Not Know Khaleda, Do Not Go to Hasina. We Do Not Want the Sick Government, trans. http://www.prothom-alo.com/bangladesh/article #Scene_2, accessed February 3, 2014.

Rahman, M. 2016: Ganges Flow in Country Lowest Since '96 Treaty. http://newagebd.net/218180/ganges-flow-in-country-lowest-since-96-treaty/, accessed April 19, 2016.

Rangan, H. 2004. From Chipko to Uttaranchal: Development, Environment, and Social Protest in the Garhwal Himalayas. In *Liberation Ecologies: Environment, Development, Social Movements.* R. Peet and M. Watts eds. Pp. 205–226. New York: Routledge.

Rashid, M. 2013. The Epic Rescue Closes. http://www.newagebd.com/detail.php?date=2013-05-14&nid=49125#.U695lu_QflU, accessed June 1, 2013.

Rehman, S. 2013. How and When Awami Government Will See off (3): 448 Killed at Rana Plaza, 2438 Rescued and Many Unattended, trans. http://www.amardeshonline.com/pages/details/2013/05/02/198437#.U695Pe_QflU, accessed June 23, 2013.

Saky, R. 2012. Kushtia Bomb Attack Kills Woman. http://www.daily-sun.com/details_yes_17–12–2012_Kushtia-Bomb-attack-kills-woman_351_1_0_3_24.html, accessed January 7, 2013.

Scott, J. C. 1979. *The Moral Economy of the Peasant: Rebellion and Subsistence in Southeast Asia.* Yale: Yale University Press.

—— 1985. *Weapons of the Weak: Everyday Forms of Peasant Resistance.* New Haven and London: Yale University Press.

—— 1990. *Domination and the Arts of Resistance: Hidden Transcripts.* New Haven, CT: Yale University Press.

Sen, A. 2008. Violence, Identity and Poverty. *Journal of Peace Research* 45(1): 5–15.

Swain, A. 1996. Displacing the Conflict: Environmental Destruction in Bangladesh and Ethnic Conflict in India. *Journal of Peace Research* 33(2): 189–204.

Times of India 2013. India Backs Shahbagh Protest. http://timesofindia.indiatimes.com/india/india-backs-shahbagh-protest/articleshow/18703429.cms, accessed March 3, 2014.

T. J. 2013. Spending in Bangladesh: The Most Bucks for the Biggest Bang. http://www.economist.com/blogs/banyan/2013/02/spending-bangladesh, accessed January 23, 2014.

Turner, T. 1997. Human Rights, Human Difference: Anthropology's Contribution to an Emancipatory Cultural Politics. *Journal of Anthropological Research* 53(3): 273–291.

UBINIG 2011. Struggle Conserving Local Rice Varieties. http://www.ubinig.org/index.php/home/showAerticle/28/english, accessed August 11, 2013.

UNBconnect 2013. Whip Noor-e-Alam Chy's Income Increases by 32,985pc in 5 Yrs: Sujan. http://unbconnect.com/candidate-income/#&panel1–2, accessed January 23, 2014.

——— 2014. Man Held with Firearm, Bombs in Kushtia. http://unbconnect.com/bomb-arrest-17/#&panel1–8, accessed June 21, 2014.

United Nations (UN) 1948. Universal Declaration of Human Rights. Available at http://www.un.org/en/universal-declaration-human-rights/. Accessed January 1, 2014.

Unnayan Onneshan 2006. *The Development Disaster: Waterlogging in the Southwest Region of Bangladesh*. Dhaka: Unnayan Onneshan.

8 Water governance at the Ganges Basin

In the previous chapters in this book, I document current water management practices in Bangladesh and analyze those practices in relationship to the agricultural system, livelihood challenges and human rights. My fieldwork evidence finds out that improved water management at the Ganges Basin, state and local level, to ensure proper seasonal flow in this region are essential to the realization of human rights to adequate food, employment, education, health and housing. In this chapter, I first describe the legal and theoretical framework for defining the right to water that I believe is best suited to my research argument. I then review the nature and severity of the problems I have documented and then summarize my understanding of their immediate and underlying causes. Finally I propose a set of recommendations for institutional reform and discuss the practicality of those recommendations in relationship to the political factors that have so far prevented the establishment of a more just and sustainable water governance system.

The human right to water

As I indicated in chapter one, there has been a significant amount of disagreement among scholars about how to define the concept of water as a human right and how best to implement it. Bakker (2012) is especially prominent among those who have criticized the water-as-a-human-right campaign as it has been promoted by anti-privatization water movements. Bakker (2012:28) argues that some groups of anti-privatization campaigners, who support water as a human right, have caused some strategic mistakes: "conflating human rights and property rights; failing to distinguish between different types of property rights and service delivery models; and thereby failing to foreclose the possibility of increasing private sector involvement in water supply." She also argues that the principle of water as a human right is "individualistic," anthropocentric" and "state-centric" (Bakker 2012:35). Based on this human rights definition, Bakker (2012:25) proposes an "alter-globalization" approach emphasizing de-commodification through management of water as a commons. She also proposes a "two-fold" approach according to which "local models of governance" should be fostered but equal emphasis should be placed on state governance reform.

An examination of the literature does demonstrate that certain attempts to legally enforce the principle of water as a human right have met with failure. This argument is routinely made in reference to the Mazibuko case in Soweto, South Africa (Bond 2012; Clark 2012) and Bustamente et al. (2012) have argued that the human right to water, as implemented in Bolivia, has given unacceptable levels of water management authority to the state, in opposition to the rights of rural communities. Many other scholars, however, have described situations in which the right to water principle has been successfully applied. Bywater (2012:207), for instance, based on evidence from India notes that local activists in Plachimada were able to shut down a Coca-Cola plant and activists in Delhi were able to prevent water utility privatization based on water as a human right argument.

Bakker's (2012:25) assertion that "human rights is a legal category applicable to individuals" is true in many instances but the concept can be applied to communities and nations as well. In my approach I rely more on the definitions of Bywater (2012), Clark (2012), Giglioli (2012), Irujo (2007), Linton (2012), Mitchell (2012) and the United Nations (1948, 2002) that emphasize collective rights. Linton (2012) emphasizes the importance of collective decision-making at all levels of a governance system. He also emphasizes the "relational" qualities of water rights: "Building on the idea that a right constitutes a relation, we might consider the right to water as involving, on one hand, a collective identity of human being as species-being (Marx 1978 [1843]:43) and on the other hand the identity of water as a process rather than a quantity" (Linton 2012:48). In her analysis of the Mazibuko case in South Africa, in which the constitutional court repudiated the human rights arguments of the plaintiffs, Clark (2012) argues that community control of water resources should be fundamental to how the right to water is defined. However, as with Bakker, Clark, Linton and many other prominent scholars in the water rights debate develop their arguments mainly in relation to urban communities and fail to recognize the very different historical factors at play in rural settings like Chapra.

My argument is consistent with those authors who see no contradiction between the idea of water as a commons and the idea of water as a human right. In this book I extend the argument of water rights still further, however, and argue that the right to water for the community of Chapra includes the right to a sufficient flow of water through local rivers, creeks and wetlands to maintain the ecological integrity of the basin and the ecological resources on which the majority of households rely. My point is that river water and related common ecological resources are not commodities; they are historically developed community property at Chapra as I demonstrate in chapters two, three and four.

It is also the case in Chapra that access to water-dependent ecological resources provides the foundation for other human rights like food, employment, education, health care and housing. The uninterrupted flow of water between international rivers, wetlands and local croplands provides essential seasonal water-borne wild vegetables like marsh herb, wild fisheries like carp, and natural fertilizers like siltation. These ecological resources are helpful for nurturing domestic animals like cattle and ducks. I read the 2002 UN human right declaration together with the 1997 UN Convention on International Watercourses and the 2004 Berlin Rules which explicitly include provisions about ecological integrity. In the case of Chapra, the definition of water as a human right is inseparable therefore from the idea of community rights, the protection of ecological integrity, and the capacity to realize other human rights which, in this environment, are heavily water-dependent.

Although the theoretical analyses of the authors noted above contribute to my argument, my fieldwork data thus opens a new avenue for understanding and defining the right to water. My evidence shows that a secure Ganges Basin flow is the prerequisite for protecting the ecological integrity of the region, ensuring the continuation of historical seasonal patterns and the availability of common natural resources and, ultimately, the ability of Chapra households to exercise their historical right to water and thereby realize other human rights. Before outlining my recommendations for how this might be achieved, I first review the evidence I have presented in this book regarding the problems occurring at Chapra and their causes.

Problems found in Chapra village

The first step towards understanding water management outcomes at Chapra was to observe community practices and identify benefits and problems. My field work revealed that ecosystem degradation is a major cause of local livelihood challenges. Unpredictability of water supply, continuous and extreme water shortages and occasional severe flooding are major aspects of this degradation. Because of the Ganges Basin flow reduction, local livelihood patterns are changing and people's traditional water sources, like wetlands, canals and oxbow lakes, are no longer available for drinking, household and agricultural activities. The disappearance of local water bodies is causing the loss of ecosystem services such as cropland siltation, river fisheries and wild vegetables. The loss of common pool resources is weakening the socioeconomic base of the agro-ecological economy that, in turn, disrupts community livelihoods and bonds. Well-off households are less likely than they once were to share their surplus croplands, seeds, fruits and fish with less well-off neighbors.

In addition to the loss of essential ecological resources, the traditional six seasons of khora, borsha, sharat, hemanta, sheet and bosonta in Bangladesh are no longer recognizable and only two major seasons of khora and borsha are visible in the local area. Secondly, even within the khora and borsha seasons, water extremes of bonna and drought are becoming dominant. In some years, a regular borsha flow is not available and, afterwards, a sudden bonna appears in local communities. More alarmingly, the length of the khora season is increasing and the length of the borsha season is decreasing. During the borsha season, lower river flow causes sedimentation, creates charland and increases salinity.

These environmental changes, in turn, contribute to more extreme flooding, more river bank erosion and embankment failures and longer periods of water stagnation. These events occur throughout the year leading to the loss of household assets, infrastructure and agricultural equipment. Furthermore, these vulnerabilities destroy public utilities like schools, hospitals, roads and natural watercourses.

In addition to the loss of ecosystem services, these changes are causing a drop in crop productivity, reduced numbers of livestock, changes to cropping patterns and economic hardships associated with the commodification of seeds, land, fertilizer and water. Local ecological knowledge about higher, intermediate and lower cropland uses does not apply anymore because of these changes. Traditional patterns of employment based on local ecological knowledge are less available with negative consequences for a wide range of employment groups such as boatmen and fishermen.

These eco-agricultural changes create unequal outcomes for the rich and poor who are now increasingly marginalized and unable to meet their basic needs. Despite extreme efforts, the marginalized households are unable to secure the minimum income needed for survival and this causes health rights violations in relation to food, employment, housing, education and health. Ninety-five percent of my household survey respondents report that water access failure is responsible for agricultural and livelihood failures. Eighty-seven percent report experiencing prolonged periods of unemployment due to water shortages and ecosystem failures. Due to these unemployment, 81 percent of my survey respondents report experiencing food shortages severe enough to be categorized as food rights violations. Unemployment and food shortages, in turn, diminish peoples' access to education, health and adequate housing. Among my survey respondents, 55 percent report being unable to educate their children; 81 percent report experiencing health rights violations; 86 percent report housing rights failures. Finding no other alternatives, they live in extreme poverty in high risk areas along river banks and embankments, and encounter displacements from their ancestral homes.

Marginalized people are also experiencing severe social problems due to these human rights violations and the breakdown of normal relations of respect and reciprocity. Starved people do not hesitate, for example, to steal food from restaurant or household. Unemployed people desperately search for any type of employment opportunity, including illegal activities like vote rigging during elections. Many local farmers interfere illegally with the GK Project canals in order to get water or remove extra water depending on the circumstances. Elites control development programs at the national and local levels in their own interests, ignoring the increasingly devastating effects of their actions on the majority of people. The state itself becomes increasingly criminalized as the party in power exploits law enforcement agencies and judicial systems to retain power. Consequently, corruption and other social problems are increasing at every level of society, threatening the viability of democracy in Bangladesh.

Causes of water and agricultural challenges for Chapra people

Ecocracy

The diversion of water in India by the Farakka Barrage is the single most immediate cause of water shortages and associated hardships throughout the Ganges Dependent Area in Bangladesh, but my study indicates that the developmentalist ideologies of both India and Bangladesh are the root cause of ecosystem failures that cause water shortages and the other problems identified above. Their governance approaches are top-down, neoliberal, technocratic and ecocratic and favor elite interests over the interests of the majority. These approaches decrease the quantity of ecological resource commons and, through privatization and commoditization, reduce the access of marginalized groups to the traditional resources that remain and to the new resources created by new technologies. Networks of corruption and patronage ensure that even the best planned development projects operate less effectively than they should. From this perspective the Farakka Barrage diversion is best understood as just one of many expressions of the underlying cause with the GK Project and GRRP providing two additional examples.

The scale of the problem in Bangladesh is magnified by the involvement of powerful international agencies like the World Bank and by the massive amounts of foreign aid that flow into the country. These agencies assist the central government to undertake development initiatives that, on the surface, appear necessary to poverty reduction but that often have perverse effects. The government has followed the terms of the World Bank's Poverty Reduction Strategy Paper (International Monetary Fund 2013), has implemented the Flood Control, Irrigation and Drainage (FCDI) Plan, the Flood Action Plan (FAP), and Integrated Water Resource Management (IWRM) plans but most of these are closely linked to neoliberal economic policy as articulated by the most recent Structural Adjustment Plan (SAP) and the Millennium Development Goals. Neoliberal approaches have also long dominated the government's approach to agricultural development and trade liberalization (International Monetary Fund 2013:280). It is not my intent in this chapter to argue against neoliberal economic policy on ideological grounds but rather to point out the empirical evidence that demonstrates its failure in Chapra and similar communities throughout the Ganges Dependent Area. Its failure is most evident in respect to the outcomes of government interventions that promote privatization and commodification.

Water commodification in the Chapra region has been achieved in a variety of ways, through technocentric development approach, for instance, that removes water from natural watercourses to which people have traditional rights and diverts it into canals where rights must be purchased. Government support for the private construction of unregulated, deep tube-wells also results in the privatization and commodification of water. Currently, the average cost of irrigation water per acre of cropland in Chapra is $84 a year which, taken together with the cost of land, seeds, fertilizer, pesticides and farm equipment is beyond the means of most households like Joardar, Kafil, Malek and Faruk. Privatization and commodification do not always create perverse outcomes but the evidence demonstrates that they have in this particular setting. Continuous commercial exploitation of water for agricultural production and other purposes is not only responsible for reducing local ecological resources and biodiversity, it causes human rights concerns over the marginalized people. This exploitation is promoted throughout the Ganges Basin with the different development projects like the Farakka Barrage in India, and the GK and GRRP in Bangladesh.

Farakka Barrage, the bilateral agreement and hydropolitics

The diversion of water away from the Ganges River before it reaches Bangladesh is clearly the most important of all the direct causes of hardship for Chapra residents. Indian Government has followed a bilateral approach in respect to the Farakka Barrage and ignore the importance of China and Nepal from consideration. This bilateral approach allows India to exploit its unequal power relationship with Bangladesh, arguing against the downstream rights of Bangladesh in the case of the Ganges River Treaty, even though it insists on its own downstream rights in its agreements with Nepal and China. India is only able to maintain this position because of its economic and military dominance in the region and the fact that it is the upstream country on the eastern, western and northern borders of Bangladesh.

India's unilateral decision to construct the Farakka Barrage without regard to downstream effects and its unwillingness to negotiate multilateral agreements for the Ganges as a whole, violates international water laws, most notably the 1997 UN Watercourses Convention. Its actions violate international customary law as articulated by the 2004 Berlin Rules. Its actions also constitute a violation of the right to water of downstream communities as identified by the 1948 UN Declaration of Human Rights, the 1966 Convention on Economic, Social and Cultural Rights and General Comment 15 which updates the 1976 Convention is respect to the right to water. Table 8.1 points out that the basin communities at Chapra encounter major ecological vulnerabilities and human rights violations because of the continuous Farakka diversion.

Table 8.1 describes the Farakka diversion from the Ganges flow creates water rights concerns and major ecological vulnerabilities like flood, drought, water stagnation, river bank erosion, or embankment failure that reduce ecological resources. These vulnerabilities create major challenges for ecological resources like *borsha* flow, cropping patterns, seeds, fertilizer, fisheries, water-borne wild vegetables, domestic animals and traditional occupations. These challenges cause concerns over agricultural production and traditional employment practices for

Table 8.1 The effects of Farakka Barrage on human rights issues at Chapra

Indicators	Effects of the Farakka Diversion	Human Rights Issues
Water	Water reduction of regional and local water bodies	Water rights
	Extinction of many local water bodies	
	Flood, drought, river bank erosion, and water stagnation	
Food	Challenges of practicing agro-ecological skills	Food and nutrition rights
	Failures of regular seasonal patterns	
	Reduction of production materials	Cultural practice rights
	Extinction of water bodies reduce fish production	
	Vegetable failure from agro-ecological sources	
	Domestic animal failure due to ecological resource failures	
	Reduction of garden from ecological services	
Work	Agricultural production failures reduce the scope	Employment rights
	Reduction of ecology based crop production	
	Traditional occupational practice reduction	
Housing	No more necessary to adjust with the basin flow	Housing rights
	Connected with road based communication	Displacements
	No need to raise platform due to the basin failure	
	Housing material failure from ecological sources	
Health	Environmental pollution and capability reduction	Health care rights
	Damage and loss of health infrastructure and transportation	Education rights
Education	Capability reduction due to crop and resource failure	
	Damage and loss of health infrastructure and transportation	

Source: Author's research data

the marginalized people at Chapra. Their concerns over food and employment rights are responsible for education, health care and housing rights failures which deteriorate further with the technological domination over the water and agricultural management in Bangladesh.

Technocentric water management

India and Bangladesh both pursue technocentric water development approaches that damage rather than maintain the ecological integrity of the basin. The list of negative ecological impacts due to the combined effects of the Farakka Barrage, GK Project and GRRP is long and growing. They have altered normal seasonal patterns and climate change is likely to intensify these effects in the years to come. The GK Project blocks the normal flow of water in rivers, creeks, lakes and wetlands, redirecting it to canals that do not support the types of riparian habitat on which people and many other species depend. Flood control embankments have been built without regard to the water stagnation problems they create and the illnesses that follow. Unpredictable water flows, river bank erosions and embankment failures interfere with the ability of most households to grow subsistence crops year around. Groundwater extraction is lowering the water table and causing arsenic contamination of drinking water supplies. Many local fish species are now extinct. Local rice is now contaminated with cadmium from chemical fertilizers.

Most of these consequences are originated from the water and agricultural policy components of the government in Bangladesh. The policy components like the Gorai River Restoration Project (GRRP) create major ecological vulnerabilities which are responsible for agricultural production loss and employment reduction. The other development projects like the Ganges-Kobodak (GK) project fail to overcome the negative effects of the Farakka but they cause multiple socioecological effects on Chapra people. Table 8.2 points out that these technologies are responsible for water and agricultural commodification and social inequality that create human rights violations.

Table 8.2 points out that the government in Bangladesh introduced the different technologies based on water and agricultural policies to overcome effects of the Farakka diversion that creates further ecological and livelihood vulnerabilities. Based on this policy approach, the foreign-educated scientists describe what is knowledge or not in water and agriculture policies that deliberately reduce importance of local ecological knowledge. The government policies introduced water technologies like the GK project, Shallow Tube-Well (STW), and Deep Tube-Well (DTW) for agricultural production. The policies also introduced the HYV crops, chemical fertilizers and pesticides to overcome consequences of the Farakka diversion. However, these policy components are responsible for river ecosystem failures of borsha (wet) and khora (dry) seasons, for ecological vulnerabilities of flooding and drought, for commodification and social inequality which are responsible for marginalized people's human rights violations.

As the environmental characteristics of the region change and traditional agricultural practices become less viable, the local ecological knowledge that formerly sustained these communities is also being lost. Two new generations have been born since the Farakka Barrage was built and few of the young adults in the community today possess the sophisticated knowledge of their elders. This knowledge, as well as scientific knowledge, should inform local water

Table 8.2 Water and agriculture policies in Bangladesh and their effects on Chapra community

Indicator	Policy Components	Effects on Community	Human Rights Issues
Water	The GRRP, the GK project, DTW, and STW replace the Ganges Basin flow	Water privatization Social inequality	Water rights
Food	Local seeds and siltation replace with HYV crops and chemical fertilizer Hybrid replace local fish Hybrid vegetable increases Poultry business increases Foreign tree and fruit	Commodification of seed, fisheries, vegetable, poultry Social inequality	Food and nutrition rights Cultural practice rights
Work	Technology-based employment replaces traditional occupations	Unemployment and exploitation	Employment rights
Housing	Corrugated iron sheet, bricks, and cements replace ecological resources	Commodification	Housing rights Displacements
Health Care Education	Privatization	Commodification Social inequality	Failures of health and education rights

Source: Author's research data

management practices but, given its association today with marginalized farming households rather than with whole communities, its reincorporation into management processes grows less likely day-by-day.

Elite control of local institutions

Class inequality at Chapra can be understood as a problem in and of itself but also as one of the main causes of the hardships now being experienced by the majority of households. My evidence indicates that class inequalities have increased sharply over the past few decades and that a small group of rich farmers now has a virtual monopoly over water resources, land, capital and even the government safety-net programs, which in theory, should benefit marginalized households. These monopolies provide significant opportunities for upward social mobility as elites establish new businesses, poultry farms and fisheries and the marginalized people become their consumers. Especially striking is the extent to which the elites are able to control all levels of local governments as they are now constituted by national legislation. They control Union Councils through intimidation and corrupt election practices. They control Upazila and District Councils through the system of appointments dominated by the local MP and Union Council Chairmen. Through these mechanisms, they are able, in turn, to control the water, irrigation and market management committees. Through connections to national elites, they control the Department of Agricultural Marketing and Agribusiness Management and a host of other departments and ministries that manage essential resources. Marginalized people have very little room for raising their voices against this domination, since they now depend on these same rich people for every aspect of their livelihoods. Local agricultural materials and government subsidies are under the control of the elite families. As traditional employment opportunities decline, the poor are increasingly dependent on the elite for employment in their commercial operations.

As the total number of marginalized people increases, they face displacement and human rights violations and join the culture of protest that has now become a dominant feature of life in Bangladesh. Along with these Farakka effects, these violations are motivated and escalated by neoliberal domination over local water and agricultural management in Bangladesh. Finding no other alternatives, local people are raising their voices with different movements. The level of movements is sometimes so extreme and spatially and temporally widespread that statecraft in Bangladesh and South Asia will encounter an uncertain future if the elites continue to dismiss and misrepresent the movements. The objectives of these movements are to promote grassroots inclusion in resource management at local, national and basin-wide levels. The Ganges Basin countries need to realize the root causes of these movements to protect the government systems before running out of time. In the next and concluding section of this chapter, therefore, I outline a set of recommendations for institutional reform that could potentially help to reverse this pattern by aligning water governance practices with the broader goals of democratization and respect for human rights.

Recommendations for promoting governance approach

Four principles inform the recommendations I make in this section: (1) the current technocentric development approach needs to change to an ecocentric approach that recognizes the need to protect the ecological integrity of watersheds; (2) the current bilateral agreement between India and Bangladesh needs to be replaced by a multilateral agreement in accordance with existing international water law; (3) water management institutions at all levels, from the local to the international, need to be reformed to allow for more equitable forms of community participation and inclusion of local knowledge; and, (4) water governance institutions should be designed to protect the human rights of all citizens, rather than to promote forms of development likely to produce inherently inequitable outcomes. After a brief discussion of these four principles, I make specific recommendations calling for the creation of two new institutions, one at the international watershed level and the other at the ward level and for comprehensive reform of other existing water management institutions. I also discuss the potential for new institutions to act as catalysts for the reform of existing institutions and conclude with a discussion of the practicality of these proposals given the current political climate in the countries of the Ganges Basin.

An ecocentric water management approach with community inclusion at local, national and basin-wide levels could provide the foundation for community sustainability and human rights protection. This approach would respect local seasonal patterns, working with rather than against them, to protect local wet-lands, lakes and canals and preserve sufficient river flow to protect local common property resources like wild fish and arum roots. Protection of these ecological resources will also provide a foundation for promoting community bonds, social justice and human rights.

In order to succeed on a watershed level, an ecocentric approach will have to be multilateral rather than bilateral. There is no easy or immediate solution for the fact that India is the dominant regional power in a position to dictate the basin management with other basin countries relatively less powerful to India. In the long run, a multilateral approach at least offers the possibility of balancing out the asymmetric power relation of India and Bangladesh and limit the likelihood that India and China will carve out spheres of interest in the region that do not respect the needs of Nepal and Bangladesh (Bandyo-padhyay and Ghosh 2009). A multilateral approach also provides more oppor-tunity to develop a consistent set of principles for water-sharing, based on international water laws that address the rights of downstream countries. A multilateral approach is also more consistent with the adoption of best practices as developed in other settings, such as those that guide Integrated Water Resource Management approaches (Orlove and Caton 2010) and nested governance approaches (Ostrom 1990).

The democratization of water management institutions through community inclusion can support the historical use of local knowledge and rights to com-mon property resources. This inclusion approach must occur at union, upazila, district, national and basin-wide levels but needs to begin at the village and ward levels where people feel most secure in their ability to elect a local repre-sentative without fear of intimidation and vote-rigging. There is no easy way to generate more democratic processes at regional and national scales but informal village level practices are characterized by an inclusionary approach and therefore offer some hope as a catalyst for a wider set of reforms. More com-munity inclusion in management institutions could go a long way towards resolving the human rights issues associated with the culture of repression and resistance in the country. Those whose rights are being violated are executing "do or die" struggles and the level of political conflict and political unrest is so severe that it creates extreme risk for the effective functioning of government systems in Bangladesh.

Incorporating the human right to water as a fundamental principle of water governance could go a long way toward ending the marginalization of rural farming households and their displacement and migration to urban slums. As argued throughout this book, the right to water in rural settings like Chapra facilitates the realization of other fundamental rights, not by spending more money on safety-net programs but by allowing people to feed themselves and practice the livelihoods to which their knowledge and skills are suited.

Elected basin-wide institution

Given the ineffectiveness of the current appointed bilateral Joint River Commission, I recommend the creation of a new, elected multilateral governance institution that could serve, initially, as an advisory body to the JRC and to national governments, but which would eventually replace the JRC or acquire sufficient legal authority to veto JRC decisions with which it did not agree. The proposed basin water governance institution would be composed of community representatives elected from every Ganges Basin country, China, Nepal, India and Bangladesh. The total number of representatives from each country could be as few as 6 or as many as 10, so that the total number of members of the institution would be between 24 and 40. Every country should have equal representation, irrespective of geographic size and population, in order to limit the extent to which any one country can dominate the others.

I will not speculate here on the election processes for other countries but, in Bangladesh, each elected member should represent a region located within the Ganges Basin and voting should occur only within such regions. These representatives should be elected approximately every four years to ensure democratic processes and responsiveness to the needs of the communities, regions and nations they represent. Rather than attempt to elect these members by the popular vote of all citizens, it will be more practical to elect them from within the District Councils, thus ensuring membership from all districts in Bangladesh within the Ganges Basin Dependent Area. District Councils will first have to undergo some reforms, however, which I describe below and, until that happens, representatives could be selected initially by civil society organizations at the district level that are focused on environmental and human rights issues. Assuming that basin governments would also not initially provide financial and technical support to the institution, such support would have to be provided by civil society organizations. Inevitably, this institution would begin its work as a pressure group, incorporating community voices and communicating them to the JRC, to national governments and to the general public in an advisory capacity. Members of the governance institution would thus have to develop the parameters of their work gradually over time but ideally they would move towards implementation of a comprehensive water governance plan that would involve a balancing of mutual interests among basin countries and communities. A democratically elected institution would not be able to achieve this balance simply by being elected but this approach would at least create that possibility, as opposed to the current situation according to which appointed JRC members are entirely controlled by their central governments. In this context, soft law can work as the prerequisites for breaking down the stalemates and mistrusts existed in the Ganges Basin countries to move forward (see Abbott and Snidal 2000; Lipson 1991; Mitchell and Keilbach 2000; Mitchell and Zawahri 2015).

Given sufficient public and international support, a citizen's forum of this type might eventually acquire the legal as well as moral authority to veto the implementation of projects such as the Farakka Barrage, GK Project and GRRP that have such negative consequences for the average basin resident. Any representative could raise concerns over a specific water development program and initiate a review and process of consultation with affected local communities. The approach would thus be highly participatory, requiring representatives to consult with affected communities. The consultation process would begin at the village and ward levels and thus link to my proposal below for new water management institutions at the ward level.

There is little precedent for the creation of an institution of this type but the concept of a participatory management institution is consistent with a broad range of water governance theory, such as that emerging from the analyses of historical community rights by Bardhan (1993), multilevel distributed governance by Wagner (2012), nested institutions and common pool resource governance (Ostrom et al. 1999), community institutions (Agrawal and Gibson 2001) and, the human rights approach of Sultana and Loftus (2012). Orlove and Caton (2010) also suggest managing finite water sources by integrating different sectors and groups at regional and national levels. In order to create opportunities for community voices in water management, the current elite domination based on top-down power structures (Bandyopadhyay and Ghosh 2009) must be changed so that water governance can "scale up" from the village level (Farrington and Lobo 1997; Heyman 2004:487).

The reform of the JRC

I am recommending that the appointed Joint River Commission be replaced eventually by an elected institution but, given the current statutory authority of the JRC, this will not be practical in the immediate future. I am therefore recommending that, for now, the commission be reformed so that it includes representation from all basin countries. International water law, as stated in the Berlin Rules and the 1997 UN Convention on International Watercourses, provides excellent guidelines for multilateral and basin-wide water management approaches. Extensive documentation also exists regarding the many initiatives around the world that have incorporated these rules when setting up multilateral governance commissions (Paisley 2002). Article 4.1 of the 1997 UN Convention states that *every country* in the basin is entitled to become part of an international watercourse management system.

Consistent with the ecocentric approach recommended here, the 1997 UN Convention outlines the responsibilities of states to protect the ecological health of a shared watershed (Articles 20, 21, 22 and 23). Article 24 of the Berlin Rules specifically notes that every country has the responsibility of ensuring adequate water flow in a river basin to maintain ecological integrity. Articles 7 and 28 of the 1997 UN Convention outline the requirement "not to cause a significant harm" to basin communities and this principle is also contained in Articles 12 and 16 of the Berlin Rules. Articles 22, 23 and 32 of the Berlin Rules outline state responsibilities to overcome "extreme situations" and Article 35 provides guidelines for cooperation among basin countries to reduce the effects of drought. International water law also provides guidelines for conflict resolution and for providing compensation to injured parties. Despite the specific guidelines for securing ecological integrity and health on the 1997 UN Convention described above, it is remarkable that no legal actions have been launched, yet, in international courts to apply the rule of law in this setting. Applying the rule of law would arguably require the creation of a multilateral Joint River Commission.

A commission of this type could help to achieve the balance of power discussed above but, given the developmentalist agendas of all four countries, this balance of power is unlikely to represent the interests of marginalized communities. An elected commission, by contrast, could fulfill all the legal requirements described above, while maintaining autonomy from political interference (Nishat and Faisal 2000). In order to assist with this process, the institution should invite third party participation from an international organization, like the United Nations, that could act as monitor and play a mediating and advisory role. It will also be essential for an elected institution of this type to link its electoral and consultative processes to the grassroots level, since it is from the grassroots level that it will acquire the moral authority it needs to be a catalyst for change and institutional innovation. In the next section I attempt to provide a blueprint for how this linkage might be formalized in structural terms.

Ward-level water governance institution

As explained in chapters two and five, local level government in Bangladesh is controlled by the elites and excludes the voices of the majority of farming households. In order to encourage greater involvement by marginalized households, I am proposing the creation of formal government institutions at a lower level in the system, the ward level, where elites will find it more difficult to monopolize elected positions. Currently, there are no formal water management institutions below the level of the Upazila Council (UP) Irrigation Management Committees and the members of those committees are all bureaucrats and engineers, except for the Upazila Council Chair, who is elected and one representative farmer, who is always appointed from among the elites. The Upazila Council itself is composed of all the Union Council Chairs in the Upazila (usually 8–10 in number) and three additional members elected directly to the UP. The local elites are able to monopolize the elected positions on the UP, entirely excluding the voices of the marginalized farmers who constitute close to 92 percent of the population at my fieldwork site, Chapra. This exclusion occurs at the Union Council (UC) level as well, despite the fact that UCs are composed, in theory, of 12 elected ward representatives, one from each ward in the union. In reality, the UC members are all from elite families within their wards and gain office mainly to serve their self-interest rather than represent the interests of the ward as a whole. There is no formal ward level government in this system and no village level government either. A ward contains, in general, from two to four villages and a population of from 2 to 5 thousand.

Even though villages provide the best evidence of cooperative and inclusive institutional behavior for local water and agricultural resource management based on commons approach, they are losing their roles as soon as individualistic approach begun to dominate in natural resource management since beginning of the British Regime in Indian Subcontinent. The British established the formal boundary of this village and named as *mauza* to facilitate census and tax collection. For this system, the regime introduced private land ownership in Bengal with Permanent Settlement Act of 1793 although this land had communal rights previously (Hossen 2016). Local people are less habituated with *mauza* and more comfortable with the term *village*. Given the individualistic approach of current water management projects, the Farakka Barrage, GK project and GRRP, the village lost its significance on natural resource management. In this context, the village is often too small to either need or support their own water management institutions. Wards are both large enough to support such institutions and small enough to facilitate regular face-to-face communication among residents. It is for this reason that I am proposing the creation of a new institution at this level. Representation at the ward level, bottom level of Union Council, will need to progress upwards through the various tiers of local level government in order to further the goal of a more democratic national governance model. This pattern is already well established in the country. The UC Chairs are members of the UPs and the UP Chairs are members of the District Councils. I suggest this model should be retained but reformed and revitalized by the addition of Ward Councilors.

The governance approach I am recommending should begin then, with a Ward Council that would have direct authority over water management decisions as well as other ward level government functions. In this discussion I focus on water management, rather than attempt to define the broader mandate that would be necessary. The total number of elected ward councilors would depend on the total number of villages and the ward population but every village should have a minimum of two representatives: one male and one female. If a ward has four villages, the total number of ward members would be at least eight. Special rules would have to be developed to maximize the opportunity for equitable representation by occupational group, as well as by gender. Partisan politics should be forbidden during the election in order to avoid the political patronage networks that dominate elections at higher levels. Election Councils would need to be established in each village to oversee the election process.

Ward Councilors could be elected for a four-year term as with other government positions and each ward would need a budget sufficient, financed by local taxes and water fees, to cover the costs of managing a small office, pay the wages of at least a part-time administrator, reimburse elected members for costs incurred and provide them with modest honoraria. Elected members as well as staff should consult regularly with residents over local water concerns and issues, hold regular public meetings as well as council meetings and maintain a complete record of actions and objectives.

In order to create ward level councils the national government will necessarily have to agree to reform other local level government structures, integrate wards into the existing system and provide them with secure funding.

The reform of local level government

In order to function effectively Ward Councils would need to possess significant authority in respect to decisions now made by higher level institutions but have significant local impacts. Local water management plans would need to be developed and, once approved by ward members, should have the force of law. Higher levels of government should be required to consult with ward councils and obtain their consent before initiating programs that would undermine a local management plan.

In addition to having a clear mandate to make certain decisions autonomously at the local level, the membership of government institutions at union, upazila and district levels should be reformed to allow for the inclusion of more ward representatives. All UCs are divided into nine wards and currently each ward elects one UC member. As noted before, this individual is almost always a male member of an elite family and once elected, mainly serves his own self-interest rather than the interests of his ward as a whole. In order to avoid the total exclusion of marginalized families and women, I propose that every ward should elect two representatives to the UC, one male and one female, following the same rules as for Ward Council elections. This means that each Union Council will be composed of 24 rather than 12 elected ward representatives. Rather than directly electing the UC Chair by popular vote of all wards, as is now the case, the UC Chair should be chosen by Ward Counselors themselves after the election. Ward election councils, created in the same way as village election councils, will oversee the elections. As with the election of ward counselors, eligibility rules will need to be created to ban partisan politics, avoid voter intimidation and enable the election of non-elite members.

I also recommend an expansion of the number of UC members chosen to sit on UP Councils. Currently, all UC Chairmen are automatically UP members but I believe the representativeness of the UC will be enhanced by doubling its members. District level councils should also be reformed so that all of their members are elected rather than appointed. A final essential reform will be for the actual resource management committees created by UPs and Districts, several of which were described in chapter five, to be established as sub-committees of elected representatives rather as committees of bureaucrats and professionals appointed by their respective ministries or departments. A group of professional administrators should work with the irrigation management committee, for instance, as experts and advisors but no actions would be undertaken by them without consent from elected member who will be non-partisan. This committee needs to promote community specific understanding of the different regional concerns and issues like irrigation, flood, water stagnation, drought and erosion. Again they also need to address overlapping alignment geographically between water and catchment areas that can be described as the shifting physical drainage areas for irrigation systems and local governmental boundaries like Union Council.

Based on these issues and concerns, it is now possible to explain with more clarity how my recommendation for a new basin-wide governance institution links to my proposal for the creation of ward councils. Adding the ward council level of government lowers the center of gravity of the system and, together with the other proposed reforms, would mean that the members of district councils would be formed mainly by elected ward level representatives. District councils would thus be in an excellent position to elect from among their members the people who would serve as members of the basin-wide institution. Each country at the district level within the Ganges Basin Dependent Area would elect 6 to 10 representatives to the basin wide water governance institution. Consistent with the practices already outlined for the ward and union levels, both women and men would need to be included and measures would have to be taken to ensure as broad a representation as possible across occupational groups.

Water democracy

Institutional reform is an extremely complex process involving changes to both formal rules and informal practices. I recognize that the recommendations I am making here represent only one of many possible approaches to resolving the extreme hardships, injustices and human rights abuses now being experienced by the majority of Chapra households. But these recommendations are at least consistent with the demands of a more democratic and inclusive governance system in which community needs and the integrity of the environment are not discounted by the development aspirations of a small minority. Water governance institutions can and should be designed to protect the water rights of all people and communities following the principles articulated in UN water covenants and customary international water law. In Bangladesh and many other settings, the enforcement of international water law would support the realization of many other human rights. The Ganges Basin governance institutions need to incorporate these water-human rights principles as the philosophical base for securing agricultural productivity, employment and adequate supply of food.

I have not made specific recommendations for other Ganges Basin countries but clearly the role of participatory water governance institutions at both local and basin-wide levels needs to be recognized by India, Nepal, China and Bangladesh as well. Such institutions are essential for both community sustainability and human rights protection. Bangladesh's failure to move in this direction is creating grave concern over its ability to return from its current situation to a peaceful existence. The excessive use of force by law enforcement agencies, denial of increasing marginalization and extreme human rights violations, generate counter actions against the government and turn poor people into human bombs; they counter-attack law enforcement agencies with stones, sticks and bombs, and take extreme measures as they cease to care whether they live or die. The struggle in Bangladesh is about more than water, but it is one of many sites around the world where "water conflicts" (Chakraborty and Serageldin 2004:201) and "water wars" (Bhalla 2012) are now a prominent feature of the political landscape. This is especially tragic in Bangladesh, where the contradiction of "ample water, ample poverty," is so stark (Bandyopadhyay and Ghosh 2009:51). The basin countries need to recognize the root causes of frustration and conflict rather than defining protestors as terrorists. The governments of India and Bangladesh, in particular, need to understand the need for reform before time runs out.

Conclusion

The current water management practices on the Ganges Basin at the international, national and local levels are not suitable for protecting ecosystem and Chapra villagers' human rights. The Farakka diversion from the Ganges Basin flow creates disappearances of the regional river like the Hisna and Chitra in the Ganges Dependent Area (GDA) of Bangladesh. The disappearance of this river causes non-availability of local water bodies like *Chapra* beel for ecological services and livelihood practices. These river and ecological service failures are responsible for human rights violations. The governments in India and Bangladesh directly or indirectly causes these violations with their water development approach. The central government in India follows the unilateral or bilateral approach to the Ganges Basin management despite the basin shares with China, Nepal, and Bangladesh. This approach causes survival challenges for the marginalized basin communities in the GDA. This challenge increases further due to the top-down policy domination over water and agricultural resource management in Bangladesh. Local governments are active part of this domination that causes major economic vulnerabilities and human rights violations for local communities like Chapra. Currently, the central government promotes the top-down approach to overcome effects of the Farakka diversion. For example, the GRRP dredging seeks to restore the Gorai flow for augmenting the summer seasonal flow. The GK project was established to meet the objectives of flood control, drainage and irrigation. However, these projects fail to meet minimum objectives and cause further socioecological challenges for local communities. The marginalized people cannot afford water and agricultural technologies and face major challenges for agricultural production and employment practices. These technologies induce market economic domination and social inequality that create further human right violations of the marginalized people at Chapra. These violations do not just affect individuals alive today; they are inter-generational in impact as when families are displaced permanently from ancestral homes with almost no chance of recovering from them. These negative outcomes are responsible for causing major local movements that create insecurity and instability at local, regional, national and international levels.

The current Ganges Basin management practices need to undergo a paradigm shift from technocentric to ecocentric approaches for the goal of human rights protection. My data analyses found that the hydropolitics between India and Bangladesh, and neoliberal policy domination over water and agricultural resource management in Bangladesh are responsible for human rights violations at Chapra. The governments in India and Bangladesh need to recognize local ecosystems of *borsha* (wet) and *khora* (dry) seasons for community survivals. This recognition can protect river water rights as the prerequisites for protecting other human rights. For this human rights protection, local knowledge about the ecosystems needs to be recognized by international, national and local water governances. This is an extremely difficult choice in the current context of regional and the global ecocractic development approaches. However, it is very urgent to find out a mechanism to accept the difficult choice because of the current human rights violations and parapolitical actions create more concern over existence of marginalized people in Bangladesh. Political activism with community engagements can bring human rights organizations, development partners, grassroots groups and basin countries together.

In opposition to the current practices of the centralized and neoliberal water and agricultural management, it is very urgent to develop the bottom-up water governance systems. Multi-stakeholders" inclusion (e.g., bureaucrats, government, the World Bank, the local development organization and the local people) can perform significant role in these systems. Due to major power difference, my purpose is to make utmost priority to the local voices among these groups of people. This priority needs to be secured at every level of governance that will begin with community representatives in formulating policies at the national level. The bottom-up water and agricultural programs which include local knowledge in these policies can protect human rights. In this context, the government in Bangladesh needs to promote bottom-up agriculture and water policy formulation and implementation, and local governments need to ensure local community representation in local resource management. This governance approach can restore agro-ecological system for agricultural production and community livelihoods based on the principles of equity and ecological sustainability, as opposed to the narrower approaches that focus on water as an economic resource alone.

References

Abbott, K. W. and D. Snidal 2000. Hard and Soft Law in International Governance. *International Organization* 54(3): 421–456.

Agrawal, A. and C. Gibson 2001. *Communities and the Environment: Ethnicity, Gender, and the State in Community-based Conservations.* New Brunswick: Rutgers University Press.

Bakker, K. 2012. Commons versus Commodities: Debating the Human Right to Water. In *The Right to Water: Politics, Governance and Social Struggles.* F. Sultana and A. Loftus eds. Pp. 19–44. London and New York: Earthscan.

Bandyopadhyay, J. and N. Ghosh 2009. Holistic Engineering and Hydro-Diplomacy in the Ganges-Brahmaputra-Meghna Basin. *Economic and Political Weekly* 44(45): 50–60.

Bardhan, P. 1993. Symposium on Management of Local Commons. *Journal of Economic Perspectives* 7(4): 87–92.

Bhalla, N. 2012. Thirsty South Asia's River Rifts Threaten "Water Wars." http://in.reuters.com/article/2012/07/23/water-south-asia-war-idINDEE86M06H20 120723?feedType=RSS&feedName=globalCoverage2, accessed February 10, 2014.

Bond, P. 2012. The Right to the City and the Eco-social Commoning of Water: Discursive and Political Lessons from South Africa. In *The Right to Water: Politics, Governance and Social Struggles.* F. Sultana and A. Loftus eds. Pp. 190–206. London: Earthscan.

Bustamente R., C. Crespo and A. M. Walnycki 2012. Seeing through the Concept of Water as a Human Right in Bolivia. In *The Right to Water: Politics, Governance and Social Struggles.* F. Sultana and A. Loftus eds. Pp. 223–240. London: Earthscan.

Bywater, K. 2012. Anti-privatization Struggles and the Right to Water in India: Engendering Cultures of Opposition. In *The Right to Water: Politics, Governance and Social Struggles.* F. Sultana and A. Loftus eds. Pp. 206–222. London: Earthscan.

Chakraborty, R. and I. Serageldin 2004. Sharing of River Waters among India and Its Neighbors in the 21st Century: War or Peace? *Water International* 29(2): 201–208.

Clark, C. 2012. The Centrality of Community Participation to the Realization of the Right to Water: The Illustrative Case of South Africa. In *The Right to Water: Politics, Governance and Social Struggles.* F. Sultana and A. Loftus eds. Pp. 174–189. London: Earthscan.

Farrington, J. and C. Lobo 1997. Scaling up Participatory Watershed Development in India: Lessons from the Indo-German Watershed Development Programme. *Overseas Development Institute, 17.* London: ODI.

Giglioli, I. 2012. Rights, Citizenship and Territory: Water Politics in the West Bank. In *The Right to Water: Politics, Governance and Social Struggles.* F. Sultana and A. Loftus eds. Pp. 139–158. London: Earthscan.

Heyman, J. M. 2004. The Anthropology of Power-wielding Bureaucracies. *Human Organization* 63(4): 487–500.

International Monetary Fund (IMF) 2013. *Bangladesh: Poverty Reduction Strategy Paper (PRSP).* Washington: IMF.

Irujo, A. E. 2007. The Right to Water. *Water Resources Development* 23(2): 267–283.

Linton, J. 2012. The Human Right to Water? Water, Rights, Humans, and the Relations of Things. In *The Right to Water: Politics, Governance and Social Struggles*. F. Sultana and A. Loftus eds. Pp. 45–60. London: Earthscan.

Lipson, C. 1991. Why Are Some International Agreements Informal? *International Organization* 45(4): 495–538.

Marx, K. 1978 [1843]. In *The Marx and Engels Reader*. 2nd Edition. R. C. Tucker ed. New York: W.W. Norton & Company.

Mitchell, K. R. 2012. The Political Economy of the Right to Water: Reinvigorating the Question of Property. In *The Right to Water: Politics, Governance and Social Struggles*. F. Sultana and A. Loftus eds. Pp. 78–93. London: Earthscan.

Mitchell, R. B. and P. M. Keilbach 2000. Situation Structure and Institutional Design: Reciprocity, Coercion, and Exchange. *International Organization* 55(4): 891–917.

Mitchell, S. M. and N. A. Zawahri 2015. The Effectiveness of Treaty Design in Addressing Water Disputes. *Journal of Peace Research* 52(2): 187–200.

Nishat, A. and I. M. Faisal 2000. An Assessment of the Mechanism for Water Negotiation in the Ganges-Brahmaputra-Meghna System. *International Negotiation* 5: 289–310.

Orlove, B. and S. C. Caton 2010. Water Sustainability: Anthropological Approaches and Prospects. *The Annual Review of Anthropology* 39: 401–415.

Ostrom, E. 1990. *Governing the Commons: The Evolution of Institutions for Collective Action*. Cambridge: Cambridge University Press.

Ostrom, E., J. Burger, C. B. Field, R. B. Norgaard and D. Policansky 1999. Revisiting the Commons: Local Lessons, Global Challenges. *Science* 84, 284(5412): 278–282.

Paisley, R. 2002. Adversaries into Partners: International Water Law and the Equitable Sharing of Downstream Benefits. *Melbourne Journal of International Law* 3(2): 1–21.

Sultana, F. and A. Loftus eds. 2012. *The Right to Water: Politics, Governance and Social Struggles*. London: Earthscan.

United Nations 2002. General Comment No. 15 (2002): The Right to Water. http://www2.ohchr.org/english/issues/water/docs/CESCR_GC_15.pdf, accessed June 23, 2010.

Wagner, J. 2012. Water and the Commons Imaginary. *Current Anthropology* 53(5): 617–641.

Conclusion

This book set out to explore current water practices and challenges for the Ganges Basin communities like Chapra in the Ganges Dependent Area in Bangladesh. My fieldwork data revealed that the current water management regime causes ecocracies that result in local community's ecological resource failures. Hydropolitics between India and Bangladesh, and technocentric water management practices in Bangladesh are the root causes of these ecocracies and community survival challenges. In this context, I reviewed the key problems and their causes to arrive at recommendations for a more just water policy and governance. I concluded with recommendations for how these problems could be solved through application of human rights principles and greater community inclusion in governance processes at local, national and basin-wide levels.

The Ganges Basin communities in Bangladesh produce agricultural crops based on local ecosystem in terms of seasonal patterns of *borsha* (wet) and *khora* (dry) which are major foundation for their livelihood security and national economic development in Bangladesh. Local people get the different ecological resources like siltation, wild fish, water-borne wild vegetables, seasonal crops and fruits based on these ecosystems. Their croplands turn into wetland during the wet season and most of them get dry during the summer season. This wetland is used for rice paddy production that is the foundation for the whole year's food security as rice is the staple food in Bangladesh. After the wet season, major cropland gets dry and suitable for other crops like mustard oil or wheat production. Based on these seasonal dynamics, local people are successful in getting the different crops, fruits, domestic animals, wild fish, and vegetation which are helpful for protecting their human rights: employment, food, education, health care and housing.

However, these ecological aspects of human rights protection encounter major challenges due to hydropolitics on the Ganges Basin and neoliberal development policies in Bangladesh. Chapra villagers like other communities of the Ganges Basin are victims of this hydropolitics between India and Bangladesh; this situation is escalated by neoliberal control over local water resource management which is termed as "ecocracy." The central governments of both India and Bangladesh create this ecocracy with their top-down domination over the Ganges

Basin water management. The central government in India, for instance, established the Farakka Barrage unilaterally on the Ganges Basin, reducing flows to agricultural communities in Bangladesh at critical times in their cropping cycles. The central government in Bangladesh has been unable to restore this flow to a sufficient level and subsequently, has followed top-down water management approach. Two major examples of this approach are the Ganges-Kobodak (GK) project and the Gorai River Restoration Project (GRRP).

These development projects in Bangladesh fail to protect local communities from water and ecological vulnerabilities but they created multiple socioeconomic problems. The central and local government elites control the GRRP and the GK project for their own interests that create further socioecological vulnerabilities for the marginalized basin communities at Chapra. The dredged river needs regular basin flow to develop the ecosystem but the continuous Farakka diversion disturbs this system. Currently, sedimentations and charlands are increasing concerns over the Gorai River at Chapra. Therefore, there are no scopes for getting rainy, winter and summer seasons in coordination with *kharif*-2, *robi* and *kharif*-1 cropping patterns. Because of this ecosystem failure, farmers fail to produce crops, boatmen fail to transport passengers and fishermen fail to catch fish; all of which, in turn, create livelihood challenges and human rights violations.

The Gorai River failure causes major grounds of searching for alternative water sources for agricultural production. The GK project is one major source for Chapra people and the project has some initial success in providing water supply to croplands. However this success encounters challenges later on due to the continuous Farakka diversion from the Ganges Basin. Moreover, the project fails to develop proper water supply systems in coordination with agro-ecological system for agricultural production. The project authority fails to develop good governance that generates further challenges for local water management. It transforms local agricultural and employment practices from traditional to technocentric system that results failures of self-sufficiency and domination of globalization on the grass root communities at Chapra. The project displaces egalitarian water practices and causes farmers' hierarchies and bureaucratic complexities. The rich farmers are successful in promoting commercial agricultural production and agri-business while the marginalized farmers are encountering survival challenges and facing human rights violations. Furthermore, the project created displacements and changed local settlement patterns with major infrastructures like canals, buildings and roads which blocked local water bodies like the Kaligangya River.

In coordination with the GK project, the government introduced HYV crops, chemical fertilizers and pesticides for ensuring agricultural production which are driving out local ecosystems and local knowledge. Chemical fertilizers and pesticides drive out the ecology based fertilizers like siltation. Local production materials like *langol* are replaced with plow machines. Moreover, local winnow fans are replaced with wheeler machines. Currently, the basin communities do not have scopes for accessing river water, local seeds and ecological fertilizers that cause major agricultural production challenges.

The central government introduced some safety-net programs to overcome these challenges and local governments implement these programs with the different management committees at the district, upazila and union levels that follow the top-down approach. The majority members of the committees are government staffs who are controlled by the superior officials of the UP and UC based on "blessings" of local MPs. They are major decision makers to distribute government benefits of water, seed, chemical fertilizer, education, health care and housing. They make lists of local marginalized people for the safety-net program benefits and exploit the systems for their own interests.

The marginalized groups of people are excluded from these committees. For example, the marginalized people do not have representation in the upazila employment management committee although they are 92 percent of the people in Kushtia. The 12-member irrigation management committee at the upazila level includes a single farmer from 270,000 people in Kumarlhali Upazila. This exclusion causes gaps between community voices and safety-net programs. For example, local communities want ecosystem-based water and agricultural programs but the central government establishes the market economic system based programs. This gap creates further livelihood challenges and human rights concerns. The decreases of agricultural production and employment increase these challenges for Chapra people's family spending; an average household needs US$160 monthly for food and well-being items but 92 percent of people, like Billal and Joardar, cannot earn half of this amount. Consequently, they encounter food, education, health care and housing rights failures. Furthermore, the extremely marginalized groups of people like physically disabled, widowed, elderly and chronically sick confront the worst human rights failures. The violations of these human rights escalate further in the context of continuous ecological vulnerabilities of flooding, river bank erosion, embankment failure, water stagnation and drought.

These human rights violations displace many of the marginalized people at Chapra which is one major reason for local movements. They want to get ecosystems back by overcoming the Farakka domination and neoliberal domination over water and agricultural management. Their grievance turns into local movements when they lose livelihood assets like field and stored crops, production materials, cropland and household infrastructures because of the development projects external to them. Their level of sufferings is so acute they cannot distinguish between life and death and this generates ecological movements at the local, regional national and international levels that can also be termed as "pluralistic grassroots movements."

Some people seek to overcome economic vulnerabilities, restore the Ganges Basin ecosystem and to protect human rights. Some organizations like the International Farakka Committee are raising local voices at the international levels. Other organizations like *Noyakkrishi Anodolon* (New Farmers Movement) are working to preserve local ecosystems and biodiversity for agricultural production based on the supports of small and marginalized farming communities. However, most of these organizations are centered in the capital city, Dhaka, and their activities are limited to major cities and urban areas. Many of these civil society organizations are operated by bourgeoisie based on their individual or corporate self-interest.

Based on strategic partnership with these organizations, local people want to establish "true democracy" in the government systems and political organizations so that they can be represented in political leaderships and government systems. As a part of their movements, they do not want to tolerate the political leaders who control local development and safety-net programs for their personal interests. Their starvation and malnutrition turn them into extremely desperate people who do not care to sacrifice life for making their voices heard. This level of desperations creates major insecurity to government system and peaceful environment at local, national and international levels.

The increasing level of livelihood sufferings and human rights violations is responsible for desperate level of local movements. Local people began to make their voices heard at local and national governments in Bangladesh although the government termed them as troublemakers and applied legal action without recognizing the root causes of the problems. This government approach creates new front of movements against the repression and domination. Protestors start counter-action aggressively, and their reactions are increasing at regional and international levels because of continuous displacements and forced migration. The government in Bangladesh and India need to realize the real causes of these movements. Otherwise, they need to take responsibilities the consequences of these movements.

To recognize the root cause, the basin countries need to develop ecosystem-based water development policies free from hydropolitics and corporate control. For this ecocentric development approach, it is important to secure ecological integrity that can ensure human rights and social justice for the marginalized basin communities like Chapra. This approach can develop political tradition with the long term objective of natural tradition or *Naturpolitik* (Latour 2004:30). However, it is not an easy task to make this bridge in the context of current reality because every epistemological question is a question of political economy and political ecology. Based on this context, Latour (2004:221) emphasized on democratic institutional development to use scientific knowledge for community empowerment. The process requires better engagements of social sciences, sociology and anthropology, for freeing political ecology from corporate control (Kottak 1999:27; Van Eijck and Roth 2007:933). The non-Western anthropologists with their organic view can promote this freedom (Latour 2004:42–43). For this purpose, Latour (2004:228) points out, "the world is young, the sciences are recent, history has barely begun, and as for ecology, it is barely in its infancy: why should we have finished exploring the institutions of public life?"

As effort to explore better institutional arrangements, the current Ganges Basin management practices need to undergo a paradigm shift from technocentric to ecocentric approach for human rights protection. Water governance approach at the three major levels: (i) the Ganges Basin-wide institution, (ii) the central governments, and (iii) local governments can restore the basin ecosystems and can protect human rights. This approach can develop based on the guidelines of the international water laws, 1967 Helsinki Rules, 1997 UN Convention and 2004 Berlin Rules, and UN human rights principles, 1948 and 2002. The Berlin Rules provided guidelines for protecting "ecological integrity" by "establishing basin wide or other joint management arrangement." The basin-wide water governance institution based on the inclusion of China, Nepal, India and Bangladesh can be the major foundation for Ganges-Brahmaputra Basin ecological integrity and community human rights protection. The 1997 UN Convention argues that every basin country is entitled to be a part of an international watercourse management that can apply in the Ganges Basin governance. This convention describes guidelines for "not to cause significant harm." These guidelines can also be major foundation for the Ganges Basin governance and overcome effects of the Farakka diversion. The meaningful participation of a third party based on the international water laws and conventions can monitor the performance of water governance institution.

The government in India needs to recognize the importance of this basin-wide water governance for human rights protection. The government of Bangladesh also needs to recognize the importance of bottom-up water and agricultural resource management in coordination with this basin governance system for this purpose. Otherwise, they will be responsible for human rights violations and face the consequences of movements at the local, regional, national and international levels.

References

Hossen, M.A. 2016. Participatory Mapping for Community Empowerment. *Asian Geographer*, 33: 1–17.

Kottak, C. P. 1999. The New Ecological Anthropology. *American Anthropologist* 101(1): 23–35.

Latour, B. 2004. *Politics of Nature: How to Bring the Sciences into Democracy*. C. Porter, trans. Cambridge and London: Harvard University Press.

United Nations (UN) 1948. Universal Declaration of Human Rights. Available at http://www.un.org/en/universal-declaration-human-rights/. Accessed January 1, 2014.

Van Eijck, M. and W. Roth 2007. Keeping the Local: Recalibrating the Status of Science and Traditional Ecological Knowledge (TEK) in Education. *Science Education* 91(6): 926–947.

Index

Printed and bound by CPI Group (UK) Ltd, Croydon, CR0 4YY

21/10/2024

01777087-0014